T0229611

JOHN H. MERCER 1922-1987

expedition almost succeeded, but bad weather and a
sick colleague prevented a successful crossing.
Later, in the 1950's he participated in Eric
Shipton's successful expeditions to the South
Patagonian icefield. These early experiences
undoubtedly were fundamentally important in
stimulating his interest in the climatic history of
the region. During these trips he made enduring
friendships with the "estancieros" who referred to
him as their "Señor Johnny".

John's discovery of Feruglio's descriptions of
the interbedded glacial and lava flows on Cerro
Fraile led to one of his most important

2

accomplishments in South America. By applying the most up-to-date dating techniques to the older glacial deposits of the mesetas, he was able to demonstrate a history of glaciation extending back to the Late Miocene.

John was also interested in more recent glacial history. The Lake Region of Chile was his favourite area. "El Alto" was a familiar figure to the residents of Puerto Varas. In reference to his studies in the Lake Region of Chile, Steve Porter commented that John had an uncanny knack for discovering important sites. John is credited with having compiled a glacial chronology for the latest Quaternary history. He provided dates for the major events: the maximum of the last glaciation, the last major readvance, and the initiation of deglaciation. He remained a firm non-believer in the existence of glacial and climatic events that were equivalent in timing to the Younger Dryas Stade of the North Atlantic.

John made pioneering glacial studies in the Chilean Channels during which he established a chronology for neoglaciations which is the standard for South America. John also conducted studies on the Quelccaya icecap of Peru and made important contributions to its recent glacial history.

He also studied glacial geology in Alaska, the Canadian Arctic, Greenland, New Zealand, and Antarctica. I once asked him if he had ever felt guilty about devoting his life to something so esoteric as climate studies. He replied that he had and, as a consequence, had spent two years in Samoa surveying the island's population.

Those of us who were fortunate to study with John knew him as a shy, gentle man who detested arrogance and pomposity. Above all he was a field scientist with a love for the simple things in life: a night under the Patagonian stars; the crackle of a calafate camp fire; the smell of burnt mutton; and the warmth afforded by a glass or two of wine from a penguin jug.

Like many others I was introduced to South America field studies by John Mercer. He was a stimulating and enjoyable field companion whose friendship I will deeply miss.

Dr. John Mercer's publications on the Cenozoic Glacial History of the Southern Hemisphere.

1960   Outline of glaciological research in the Antarctic prior to   the IGY, **Glac. Notes**, 3.
1962 Glacier variations in the Antarctic, **Glaciol. Notes**, 11.
1962 Glacier variations in New Zealand, **Glaciol. Notes**, 12.
1962   Glacier variations in the Andes, **Glaciol. Notes**, 12.
1963   Glacial geology of the Ohio Range, central Horlick   Mountains, Antarctica, **Inst. Polar Studies Rept.** 8.
1965   Glacier variations in southern Patagonia, **Geog. Rev.** 55, 390-413.
1967 Glaciers of the Antarctic, **Amer. Geog. Soc. Antarctic Map   Folio Series**, 7, 10p.
1967 **Southern hemisphere glacier atlas**, U.S. Army, Natick Lab., Tech. Rept. 67-76-ES, 325.
1968 Variations of some Patagonian glaciers since the Late-Glacial, **Amer. J. Sci.**, 266, 91-109.
1968   Glacial geology of the Reedy Glacier area, Antarctica, **Geol. Soc. Amer. Bull.** 79, 471-486.
1968   The discontinuous glacio-eustatic fall in Tertiary sea   level, **Palaeogeog. Palaeoclim. Palaeoecol.** 5, 77-85.
1968   Antarctic ice and Sangamon sea level, **Internat. Assoc. Sci.   Hydrology Publ.** 79, Gen. Assembly of Berne, 1967,   Commission on Snow and Ice, p.217-225.
1969 The Allerod Oscillation: a European climatic anomaly? **Arctic and Alpine Res.** 1, 227-234.
1969   Glaciation in southern Argentina more than 2 million years   ago, **Science** 164, 823-825.
1970   Antarctic ice and interglacial sea levels, **Science** 168, 1605-1606.
1970   A former ice sheet in the Arctic Ocean? **Palaeogeog. Palaeoclim. Palaeoecol.** 8,19-27.
1970   Variations of some Patagonian glaciers since the Late- Glacial II, **Amer. J. Sci.** 269, 1-25.
1971   Cold glaciers in the central Transantarctic Mountains, Antarctica; dry ablation areas and subglacial erosion, **J. Glaciol.**, 10, 319-321.
1972   Chilean glacial chronology 20,000-11,000 carbon-14 years ago; some global comparisons, **Science** 176, 1118-1120.
1972   The lower boundary of the Holocene, **Quaternary Res.** 2, 15-24.
1972   Some observations on the glacial geology of

the Beardmore Glacier area: i, Antarctic Geology and Geophysics, R.J. Adie, Ed., Universitetsforlaget, Oslo, 427-433.

1972 Fleck, R.J., Mercer, J.H., Nairn, A.E.M., and Peterson, D.N. Chronology of Late Pliocene and Early Pleistocene glacial and magnetic events in southern Argentina, Earth and Planetary Science Letters 16, 15-22.

1972 Mercer, J.H., Fleck, R.J., Mankinen, E.A., and Sander, W., Glaciation in southern Argentina before 3.6 m.y. ago and origin of the Patagonian gravels (Rodados Patagónicos), Abstracts, Geol. Soc. Amer. Annual Meeting, Minneapolis.

1973 Cainozoic temperature trends in the southern hemisphere: Antarctic and Andean glacial evidence, Palaeoecology of Africa and Antarctica, v.6, p.85-114.

1973 Mercer, J.H. and Laugenie, C.A., Glacier in Chile ended a major readvance about 36,000 years ago: some global comparisons, Science 182, 1017-1019.

1975 Stuiver, M.; Mercer, J.H.; and Moreno, H. Erroneous date for Chilean glacial advance, Science 187, 73-74.

1975 Mercer, J.H., Fleck, R.J., Mankinen, E.A., and Sander, W., Southern Patagonia: glacial events between 4 MY and 1 MY ago, in: Quaternary Studies (R.P. Suggate and M.M. Cresswell, eds.), Royal Society of New Zealand, p.223-230.

1975 Mercer, J.H., Thompson, L.G., Marangunic, C. and Ricker, J. Peru´s Quelccaya Ice Cap: glaciological and glacial geological studies, 1974, Antarctic Journal of the United States 10, 19-24.

1976 Glacial history of southernmost South America, Quaternary Research 6, 125-166.

1977 Radiocarbon dating of the past glaciation in Peru, Geology 5, 600-604.

1978 West Antarctic ice sheet and CO2 greenhouse effect: a threat of disaster, Nature 271, 321-5.

1978 Age of earliest mid-latitude glaciation, Nature 274, 926.

1978 Glacial development and temperature trends in the Antarctic and in South America, in: Antarctic glaciation and world palaeoenvironments, E.M. van Zinderen Bakker, ed., Balkema, Amsterdam.

5

1981   Tertiary tillites of the Ross Ice Shelf area,
Antarctica,   in: Earth's pre-Pleistocene
glacial record, International   Geological
Correlation   Project   38:   Pre-Pleistocene
tillites, M.H. Hambrey and W.B. Harland, eds.
Cambridge   University Press. p.204-207.

1981 West Antarctic ice volume: the interplay of
sea level and   temperature, and a strandline
test for absence of the ice   sheet during
the last interglacial, in: Sea level, ice and
climatic change (Proceedings of the Canberra
Symposium, December, 1979), IAHS Publ. No
131, p.323-330.

1982   (with J.F. Sutter) Late Miocene-earliest
Pliocene   glaciation in southern Argentina:
implications for global   ice sheet history,
Palaeogeography,   **Palaeoclimatology,**
Palaeoecology, 38, 185-206.

1982   Holocene glacier fluctuations in southern
South America,   **Striae** 18, 35-40.

1983   Cenozoic glaciation in the Southern
Hemisphere, Ann. Rev. Earth Planetary Sci.   11,
99-132.

1983 Webb, P.N., Harwood, D.M., McKelvey, B.C.,
Mercer, J.H. and   Stott, L.D. Neogene and
older   Cenozoic   microfossils   in   high
elevation   deposits   of   the   Transantarctic
Mountains: Evidence   for   marine
sedimentation and ice volume variation   on   the
east antarctic craton, **Antarctic Journal of the**
United States, 18(5):96-97.

1984   Simultaneous   climatic   change   in   both
hemispheres   and   similar   bipolar
interglacial   warming: evidence   and
implications, in, **Climate Processes and Climate**
Sensitivity   (Maurice   Ewing   Symposium
Series, Volume 5), American Geophysical Union.

1984 Late Cainozoic glacial variations in South
America south of   the Equator, in, **Late**
Cainozoic Palaeoclimates of the   **Southern**
Hemisphere, (Proceedings of Symposium held in
Swaziland, August 1983), South African Society
for   Quaternary Research, Pretoria.

1984 Webb, P.N., Harwood, D.M., McKelvey, B.C.,
Mercer, J.H., and   Stott, L.D., Cenozoic
marine   sedimentation   and   ice-volume
variation on the East Antarctic craton, **Geology,**
12:287-291.

1985 When did open-marine conditions last prevail
in the Wilkes   and Pensacola basins, East

Antarctica, and when was the Sirius Formation emplaced? South African Journal of Science 81:243-245.

1985 Changes in the ice cover of temperate and tropical South America during the last 25,000 years. Abl. Geol. Palaont. Teil 1, 11/12:1661-1665.

1986 Southernmost Chile: a modern analog of the southern shores of the Ross embayment during Pliocene warm intervals. Antarct. J. of the U.S. (submitted).

1986 Southern Chile: contrasts in behavior of two tidewater glaciers. J. of Glaciology. (submitted).

W.J.WAYNE
Department of Geology, University of Nebraska, Lincoln, Nebr., USA

2

The diamictons of Río Blanco basin, Cordón del Plata, Mendoza

ABSTRACT

Both the sediments and the geomorphic setting of the diamictons
along Río Blanco on the piedmont east of the Cordón del Plata
indicate that they could not have been deposited there by a
glacier tongue, as I and several earlier observers have suggested
Rather, they must have been deposits of long-runout debris flows,
as argued by Polanski. Below 2600m, which is
the altitude of the most extensive moraines of the last major
glaciation, a late Pleistocene debris-flow fan fills the valley.
From 2,200 m to the mountain front the narrow, V-shaped valley
could not have been scoured by glacial ice. The debris flows
that deposited these piedmont diamictons incorporated glacially
transported clasts, so some of the flows recognized along the
Río Blanco trench may have taken place when glacier ice lay in
the cirques and upper valleys. Others, however, probably are
unrelated to glacial events.

RESUMEN

Un nuevo examen de los diamictos y su correspondiente geomorfología en la parte pedemontana al este del Cordón del Plata indican que ellos no pudieron haber sido depositados por una lengua glacial, como lo han sugerido mis estudios anteriores y los
de varios investigadores. En contraposición, esos diamictos son
depósitos de flujos de escombros, como lo sugirió Polanski. Ras
gos litológicos indican que las corrientes de escombros que de-
positaron los diamictos comenzaron en o pasaron pos una de las
morenas que fueron transportadas por los glaciares. Algunos de
los flujos de escombros pudieron haber ocurrido durante una cla
ciación mayor, cuando una lengua

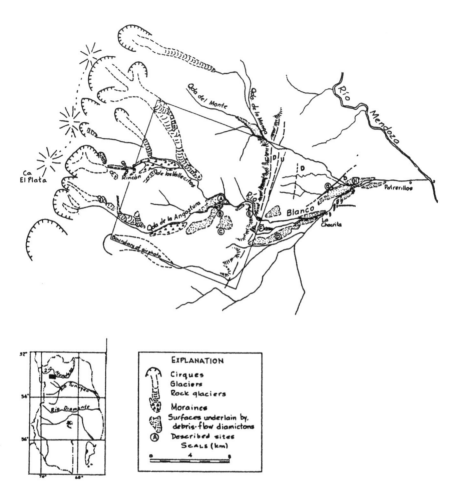

Figure 1. Distributions of diamictons in the Río Blanco basin.
Letters indicate the locations of sites mentioned in the text.
"D": down-faulted block; "U": up-faulted block. Map prepared
from NASA ERTS E-2022-13455, 13 Feb. 1975 and airphotos,
scale 1/50,000, from the Institute of Military Geography,
Argentine Army.

glacial llenaba el circo y la parte superior de la quebrada
aguas abajo del circo. Sin embargo, ninguna de las capas de
diamictos localizadas en la parte más baja de las morrenas
extremas de la última glaciación, cuya altitud alcanza 2600 m,
fueron depositadas directamente por un glaciar; es probable

que algunos de los diamictos fueron depositados por flujos
durante intervalos interglaciales. Ningún glaciar pudo haber
pasado por la quebrada, la cual es angosta y tiene una clara
forma de "V".

INTRODUCTION

Recently, Wayne and Corte (1983) and Wayne (1984) reiterated
the concept that glacial ice may have extended a considerable
distance onto the piedmont east of the Cordón del Plata during
the Pleistocene and interpreted the sediments to include
deposits of at least four glaciations. These conclusions were
based on the presence of three identifiable diamictons beyond
the outermost moraines of the last major glaciation and several
other features of these sediments that made reasonable their
identification as till. These specific features are:
  1. Their lithology, which is similar to that of unquestioned
tills higher in the valleys.
  2. The surface textures of many of the quartz grains, which
indicate glacial grinding.
  3. Presence of a few striated quartzite boulders in the
diamictons.
  4. A boulder fabric in some exposures that would have been
expected as a result of glacial deposition.
  5. Valley walls that had been scoured and polished by
glaciers larger than those that deposited the moraines of the
last major glaciation in them. In addition, the large size of
most of the cirques suggests strongly that they have been
occupied by glaciers several times, not just once.
  A more thorough study and analysis of the deposits and their
geomorphic relationships, however, has caused me to question
our earlier interpretation that these diamictons were
deposited by tongues of glacial ice that extended onto the
piedmont. Rather, I now believe it to be more likely that all
the diamictons below and beyond the most extensive moraines
of the latest or Vallecitos (=Wisconsin/Würm) glaciation
(Figure 1), about 2600 m in the basin of Río Blanco, were
deposited as debris flows, as suggested by Polanski (1953,
1962, 1966). It is my purpose in this report to clarify and
modify, where necessary, the interpretations presented earlier.

CHARACTERISTICS OF DEBRIS FLOWS AND THEIR SEDIMENTS

Although the importance of mudflow and debris-flow sediments
in the accumulation of alluvial fans on the piedmont of semiarid mountains has long been recognized (Blackwelder, 1928;
Sharp and Nobles, 1953; Polanski, 1953, 1962, 1966; Beaty,
1963, 1974), both the process and the deposits have been
investigated more thoroughly only in recent years (Lindsay,
1968; Fisher, 1971; Hsü, 1975, 1978; Innes, 1983; Costa and
Williams, 1984). Many diamictons that result from debris flow
are difficult to distinguish from ice-laid till; debris flows
are as capable as glaciers of transporting huge blocks of rock
for long distances. Debris flows generally contain large
amounts of rock flour (silt and sand-sized particles) but
little clay. The final sediment produced may range from
clast-supported, with insufficient fine material to fill the
interstices, to a matrix-supported material with scattered
clasts.

Because diamictons produced by mudflow or debris flow and
those deposited by glacial ice often appear indistinguishable
(Dreimanis and Lundqvist, 1984), several recent efforts have
been made to find means to differentiate them. Landim and
Frakes (1968) found that, although some overlap existed,
statistical evaluation of clay-to granule-sized particles
generally permitted these different sediments to be identified.
Most fabric studies of diamictons have involved ice-laid
sediments, and glacial geologists have generally thought of
mudflows as developing either weak fabrics or none. Lindsay
(1968) and Mills (1984), however, demonstrated that mudflows
and debris flows also develop fabrics, and that under some
conditions the fabric is similar to that in a till. Fisher
(1971) showed that some mudflow/debris flow deposits show
inverse grading--that is, the coarsest clasts are near the
top. Inverse grading, however, is not particularly evident
in many exposures of debris-flow diamictons.

In review papers, Polanski (1966) and, more recently, Innes
(1983) pointed out that confusion has long existed in the
terminology of the materials deposited by debris flows and
summarized the conditions reported to cause them to take
place. They also indicated that much work remains to fully
understand debris flows and their deposits. In this report,
I have used the term diamicton, which was introduced by Flint,
Sanders, and Rogers (1960a, b) as a nongenetic term to apply

to mixtures of unsorted debris with a range of clast sizes
from clay to large boulders. Most of the other terms available
for sediments of this type have genetic implications (till,
till-like), are inappropiate (fanglomerate), or awkward (pebbly
mudstone), although a term proposed by Harrington (1946),
cenoglomerate, appears to be equally nongenetic and was the
one preferred by Polanski (1966).

Diamictons deposited by debris flows are likely to change in
composition from their distal to proximal zones. The lead or
distal part of a debris flow generally is highly charged with
large clasts, but farther upstream, where larger amounts of
water become mixed with the sediment, clast sizes generally
become smaller and the frequency of large fragments diminishes.
It may be difficult, therefore, to identify the diamicton
resulting from a particular debris flow along a valley unless
outcrops are virtually continuous. Late phases of debris flows
become muddy stream flow, and often anastamosing channels are
cut into the top of the diamicton deposited in the early part
of the episode by the surges of floodwater that follow debris
deposition (Beaty, 1963). See also Pierson (1986).

Debris flows, like all flows, follow existing channels,
filling them with sediment and spilling over onto an adjacent
valley flat only if the sediment volume is sufficiently great
or if boulders provide temporary blockage (Beaty, 1963). Both
in and out of channels, the frontal margin of most bouldery
debris flows is blunt and lobate. Mudflows and debris flows
of small magnitude are fairly common, but those capable of
moving long distances involve large volumes of debris (Hsü,
1975) and take place much less frequently. Large ones involve
immense masses of rock debris that move rapidly down a steep
mountain slope, generating so much kinetic energy that motion
may continue for distances of several kilometers through
foothill valleys and across the more gentle piedmont slopes.
Debris flows may be wet or they may be nearly dry. Freshly
deposited debris-flow sediments may be remobilized by
subsequent surges and so continue downslope or down valley
after they have stopped one or more times.

Large scale debris flows--including those with which we
are concerned in this report--may originate in several ways
(Polanski, 1966; Eisbacher, 1982). Commonly, they result from
supersaturation of an unconsolidated or weathered mantle on
steep slopes, either as a result of intense storms (Beaty,

1963) or from the rapid melting of a thick snowpack. They may
also result from sudden outbursts of glacial meltwater that
is heavily loaded with debris or from the rupture of a moraine
or glacier dam that holds a lake (Lliboutry **et al.**, 1977;
Clague **et al.**, 1985). An additional means, and one that
involves some of the largest volumes with greatest runout, is
the rockfall-generated Sturtzstrom described by Heim in 1882
(Hsü 1975; 1978).

Debris flows that originate on weathered slopes are likely
to incorporate a large amount of weathered rock debris; those
that involve steep morainal accumulations or emerge from the
margin of a glacier should include a high proportion of fresh
rock material. Huge rockfalls from a mountain cliff may result
in great masses of both weathered and fresh debris. Debris
flows take place periodically, spasmodically, catastrophically.
Small ones may occur regularly; large ones are relatively
infrequent and individual events may be decades or centuries
apart (Whalley, Douglas, and Jonsson, 1983). Frequencies of
very large scale, long-distance-runout debris flows are
evidently variable, some on the order of decades or less
(Plafker and Eriksen, 1978), yet others may be separated by
millenia or longer periods of time.

RIO BLANCO VALLEY DEPOSITS

Diamictons are exposed in many places along the Río Blanco
valley between the moraines in the valleys of Angostura,
Vallecitos and Potrerillos (Figure 1), and are dominated
by abundant large quartzite clasts, most of which are fresh
and unweathered. Non-quartzite clasts, mainly granites and
rhyolites, constitute 5-20% of the rocks in diamictons. Most
of the diamictons are matrix-supported, although in a few
exposures the number of cobble-to-boulder-sized clasts is
so great that the matrix scarcely fills the interstices.
The matrix contains a small amount of clay but is dominated
by silt, sand, and granules. Although most of the sand-sized
grains in the diamicton are rock (quartzite) fragments, 10
to 20% are single-crystal quartz grains, many of which have
surface textures characteristic of glacial grinding (Wayne,
1984; Wayne and Corte, 1983). At least 4 separate beds of
diamicton are exposed along Río Blanco in stream and road cuts.

A large fan-shaped and boulder-strewn accumulation separates

the channels of the two creeks, Angostura and Vallecitos, that
join to form Río Blanco (Figure 1, site A; Figures 2, 3, 4).
From just above their junction at 2200 m to the outermost
moraine about 2600 m, its average slope is 8 degrees. Both
creeks, Angostura and Vallecitos, flow along the lateral margins
of the fan where they were forced by the sediment accumulation.
Cleanly swept banks along Arroyo Angostura expose two diamictons
beneath the surface, which is covered by a network of shallow,
anastomosing channels that are bordered by levees. The uppermost
diamicton is medium gray (2.5Y 5/4), 3 to 4 m thick where it
is exposed through midfan, and is capped with a loess veneer
20 to 30 cm thick. The boulders that litter its surface are
probably a lag produced by washing shortly after deposition.
It overlies another diamicton, somethat browner (10YR 4/3),
that has remnants of a weathered zone at its top; the
lower diamicton matrix is calcareous, as are all the diamictons
exposed farther downstream along Río Blanco, but the upper one
is not. A similar fan has formed at the piedmont edge where
Quebrada de la Manga emerges from the mountain front (Figure
1). The channels cut into its surface, however, are more sharply
defined and fresher looking than those on the surface of
the head of Río Blanco.

Along its south valley wall, Arroyo Angostura has undercut
a slope in which two additional superimposed diamictons can be
examined (Figura 1, site B; Figure 2). The lowermost part of
this exposure (Wayne and Corte, 1983, figure 7; Wayne, 1984,
p. 406) is a calcareous diamicton that contains grusified
granite clasts, a strongly developed buried alfisol profile has
been developed on it. This buried paleosol is overlain by a
 thick (6m) boulder-rich diamicton that has an aridisol profile
with a 20 cm thick stage III petrocalcic horizon (Birkeland,
1984, p. 357-359) and thin silt cap. The calcium carbonate
grains in the matrix of this and the other diamictons downstream
along Río Blanco must be secondary, deposited by groundwater
movement. No carbonate rocks have been recognized in the basin
of Río Blanco.

The additional diamictons are presented beneath much higher
surfaces in this area. One caps the ridge that extends eastward
from the moraine-choked middle segment of Quebrada de la Angostura but is about 200 m above the present creek bed (Figure
1, site C). This diamicton has a thick stage IV petrocalcic
horizon and the igneous boulders in it are highly weathered.
It was deposited along the valley of Arroyo Vallecitos when it

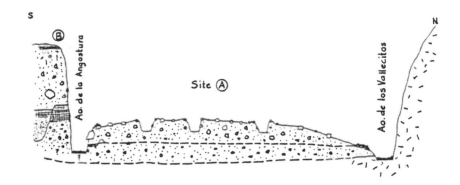

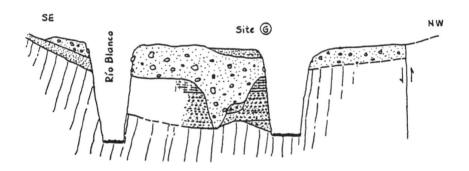

Figure 2. Diagrammatic sections across the Río Blanco valley at Sites B-A and at G.

flowed southeastward as a tributary of Río Tunuyán, before it was diverted eastward into Río Mendoza (Wayne, 1985). The other forms the cap of the Mesón del Plata just to the east of the mountain front. The soil profile on it, too, is an aridisol with a thick petrocalcic horizon; traces of shallow channels are still recognizable on the surface of the Mesón del Plata (Figure 1, site D).

Two diamictons are exposed in at least 4 places through the "narrows", a rock-walled twisting valley of Río Blanco downstream from the junction of the two creeks that form it (Figure 1, site E; 2,5). East of the front of the Cordón del Plata, Río Blanco flows through a steep-walled trench cut below the piedmont plain and exposes in several places the sediments, both diamictons and alluvial gravels, that underlie the sloping alluvial surface. Two narrow ridges of diamicton

Figure 3. Part of airphoto N° 6904B-40-5512 showing valley
characteristics between lower Vallecitos moraines and piedmont.

stand above the alluvial slope that has been cut across the
sediments filling the Río Blanco valley. The ridges are
topographically lower than the deposits of the Mesón del Plata,
yet clearly higher than those beneath the alluvial plain. The
soil profile on the one that stands as a low ridge in midvalley
near the west end of the Mesón del Plata is an aridisol with
a strongly cemented K horizon about 30 cm thick (Figure 1,
site F). Both quartzite and rhyolite boulders are present
along the ridge.

Figure 4. Bouldery surface of debris flow fan upstream from
the junction of Arroyo Vallecitos and Arroyo Angostura (Site A).

    Previous workers (Corte, 1957; Polanski, 1953; Wayne and
Corte, 1983) have recognized only a single diamicton within
the valley and below the remnant that caps the Mesón del Plata.
Although only one diamicton, generally overlying and/or capped
by bouldery gravel, can be seen in most of the cleanly swept
cliffs along Río Blanco, at least 3 are present within the
sediment fill of the valley. Two of these can be seen in almost
continuous section along a small tributary that enters Río
Blanco across a concrete-surfaced ford upstream from Potrerillos (Figure 1, Site G). In this group of exposures, which
are on the downthrown side of a fault with a displacement of
a few 10's of meters, a poorly sorted thin basal gravel
(diamicton?) rests on red Tertiary sandstone. It is overlain
by a bed of coarse gravel capped by a noncalcareous red
(2.5YR 5/8) sandy silt bed about 2 m thick (Figures 2, 6). The
level top of this alluvially deposited fine-grained layer is
buried beneath a younger bouldery diamicton, olive (5Y 5/3)
in color, which not only covers the old alluvial surface but
also fills a gully that evidently had been eroded through the
beds beneath it (Figure 7). It can also be seen as a greenish-
gray layer capping the steeply dipping Tertiary rocks of
Los Mogotes Formation along the south side of the valley of

Figure 5. Two superimposed diamictons exposed in the narrows
of Río Blanco valley (Site E).

Río Blanco between La Chacrita and Potrerillos, and its lobate
margin overlaps an alluvial fan that was deposited by a small
stream emerging from the badlands south of Río Blanco.
Although rock chips are common, few boulders can be seen at the
surface. The soil profile contains a large amount of silt and
there is a Bk horizon (stage II+) at a depth of 40 to 60 cm
beneath the surface. In some places, a post-depositional stream
has planed the surface and covered it with alluvium (Figure 2).
The diamicton that caps the Tertiary sediments on the north
side of the stream at site G (Figure 2) contains a noticeable
number of strongly weathered rhyolitic clasts, in contrast to
the high abundance of quartzites in the thick diamicton exposed
along the stream. These features indicate that two different
diamictons are exposed at this site.

ORIGIN OF THE DIAMICTONS

The dominance of quartzite clasts in the diamictons along
Río Blanco suggests that most of them came from the upper part
of Quebrada de la Angostura, although the presence of rhyolitic
material means that at least some of the debris came from one

19

Figure 6. Diamicton overlying irregular surface on Tertiary sandstones, with a second and younger diamicton filling a gully cut through it (Site G).

of the Vallecitos valleys. A virtual absence of striated cobbles and boulders in the Río Blanco diamictons was cited by Polanski (1953) as a line of evidence against glacial deposition, but alpine glaciers generally produce few striated clasts. Evenson et al. (1986) have pointed out that a high proportion of the sediment attributed to valley glaciers and deposited as till has really been moved little by ice; rather, most of it is delivered by running water or rock fall to the ice, which may deposit it after minimal transport. Even though only 4 striated clasts were observed in this investigation, they along with the presence of glacially fractured quartz grains make it seem likely that the diamictons in the Río Blanco valley between the lowest moraines (about 2600 m) and Potrerillos (1400 m) were derived at least in part from glacially deposited sediments. Unfortunately, whether the debris flows that extended so far down the piedmont slope were generated from the front of an active glacier, from saturation of moraines and weathered rock debris in the upper valley by intense storms or by snow-pack melt can not be determined. Relative dating and internal stratigraphy indicate that several such events took place through a long span of time, perhaps several hundred thousand

Figure 7. Two diamictons with intervening alluvial sediments exposed in piedmont channel (Site G).

years, after erosion had cut the valleys to their maximum depth, and as the river has re-excavated the alluvial and mud-flow accumulations that have repeatedly filled its valley.

Although these diamictons were thought to be deposits of two glaciations by Wayne and Corte (1983), and the sediments do show evidence of glacial action, glacier ice could not have transported them so far from the cordillera without having greatly enlarged the valley through the "narrows", where Río Blanco leaves the cordillera. The narrowness and twisted nature of the valley through this reach (Figure 3), as well as the weathered rock surfaces of the valley walls, indicate that a glacier could not have passed through it in order to deposit the diamictons that the stream exposes along its banks. Ice would also have smoothed the irregular surface of poorly consolidated Tertiary clastic rocks upon which the diamictons lie. Corte (1957, p.13) noted the "V" shape of the Río Blanco valley, but suggested that glacier ice plugged the bottom of the valley and that the moving part of the glacier flowed across the "dead" ice plug at a higher level. It would be unusual, however, for an ice stream to behave in this way.

Three diamictons can be recognized in the high banks of Río Blanco for about 9 km east of the east margin of the

Cordillera (Figures 1, 6), where the stream reaches an altitude of 1525 m; a single one can be followed along the banks nearly to the Río Mendoza valley, at 1400 m. Debris flows with the runout distances of these could readily have moved this far, however, without eroding the underlying material. In passing through the constricted valley at the "narrows", such flows probably would have thickened sufficiently to generate a sequence of surges (Jian **et al.**, 1983) that would have kept the debris moving well out onto the piedmont. The two diamictons (Figure 5) that seem superimposed in some exposures in the narrows may, in fact, represent surges rather than two depositional events (Fisher, 1971, p.920). Material formerly transported by glacial ice surely made up a significant part of the debris-flow masses, which, for the one with the greatest run-out distance, may have exceeded 20 x 10^6 m^3 in volume.

Saturation of glacially deposited material and generation of small debris flows from the margins of the present debris-covered glaciers above 3500 m in Quebrada de la Angostura is a fairly common event. A larger-than-usual debris flow in December 1982, caused by rapid melting of an unusually heavy snowpack, nearly filled the channel downstream to 2250 m, but most of the sediment had been removed through stream erosion within a few weeks, and by December 1984, only remnants were visible along the channel sides.

One hypothesis that might explain these long runout diamictons as well as the presence of glacially-derived materials in them would be a major rockfall from the cirque headwall onto a glacier surface, perhaps seismically induced. The cirque headwall of Quebrada de la Angostura is a cliff that rises nearly 1000 m above the floor of the valley (Figure 8). Today, it is not occupied by a glacier, but during de maximum of the last glaciation, the cliff surely stood at least 500 m above the surface of the ice that filled the cirque. A major rockfall (or avalanche) onto the surface of a glacier in the head of Quebrada de la Angostura surely would have generated a debris of massive proportions. Such a rockfall would pick up snow and ice as it passed down over the glacier surface and incorporate meltwater as it followed the valley beyond the margin of the ice tongue. A debris flow of this magnitude could readily continue in motion until it reached the valley of Río Mendoza, particularly if it were following a channel through the piedmont zone. Although

Figure 8. Cerro El Plata, showing a large cirque at head of Quebrada de la Angostura.

such a hypothesis regarding the origin of one or more of the diamictons of the Río Blanco valley is speculative, it would account for all of the observed features of the sediments that suggest a glacial origin for some of the materials as well as for the geomorphic characteristics of the valley below 2600 m that make it unlikely that glacier ice ever passed through it. That rockfall/ debris-flow sediments accumulate in the central Andes in this manner was clearly demonstrated at Nevados Huascarán in 1962 and 1970 (Plafker and Ericksen, 1978; Browning, 1973); Keefer (1984) called attention to the importance of seismic events as a triggering mechanism. The piedmont fault that contributed to preservation of the diamictons near La Chacrita is evidence of Late Quaternary seismic activity in the area.

STRATIGRAPHIC-CLIMATIC INTERPRETATIONS OF THE DIAMICTONS

Soil profile characteristics, stratigraphic superposition, topographic position, morphologic changes, and degree of alteration of glacially fractured quartz grains all provide guides to establish a relative age framework for this sequence of diamictons and associated sediments. The oldest of the diamictons associated with Río Blanco lies in a fragment of a trough that trends southeastward and stands about 200 m above the present channel (Figure 1, site C). Wayne (1985) pointed out that it was deposited at a time when the headwater parts of the Río Blanco basin drained southeastward toward the valley of Río Tunuyán and that it is probably either late Pliocene or early Pleistocene in age. During an episode of structural deformation along the front of the Cordón del Plata, the part of the basin that became Río Blanco was diverted eastward into Río Mendoza and the diamicton that now caps the Mesón del Plata (Figure 1, site D) was deposited along it. Remnants of channels can be seen on the surface of the Mesón del Plata. Both groups of diamictons and their associated gravels constitute the unit Polanski (1962) named the Mesón del Plata Formation. The other diamictons preserved along the Río Blanco valley, both within the Cordón del Plata and on the piedmont to the east, were emplaced after the river had entrenched its channel to the new base level at Río Mendoza. Nearly all of these diamictons are associated with fluvial sediments and are included within Polanski's (1962, 1972) La Invernada Formation.

During glaciations, Río Mendoza flowed at a higher level than now, as evidenced by the gravel terraces along it (Brunotte, 1983). During times of glaciations in the Andes, the higher base level along Río Mendoza surely caused aggradation along Río Blanco and all other tributaries; hence it would be logical to correlate the alluvial sediments of the higher terraces with glaciations. The diamictons observed in the base of the exposures along the valley of Río Blanco may have been deposited during interglaciations, when Río Mendoza provided a lower base level for its tributaries, as it does now.

The youngest diamicton in this sequence underlies the surface of the large fan between 2600 and 2200 m that fills the valley just upstream from the junction of Arroyo de la Angostura and Arroyo de los Vallecitos (Figures 1, Site A, 2, 3). It is littered with unweathered boulders, about 10% of which are granites, and is covered by a topographically fresh network of levee-bordered channels. The loess veneer, soil profile development, and rounding of granitic clasts are comparable to those of the outermost Vallecitos moraines at 2600-2650 m in the valley at the head of the fan. It seems likely, then that this particular debris flow, which is 3 to 4 m thick and comprises at least 2×10^6 m^3, took place while an ice tongue

24

lay in the valley above it during the last (Wisconsinan) glaciation (Wayne and Corte, 1983, fig. 2).

The upper 30-40 cm of the soil profile developed on the diamicton that underlies the terrace surface just upstream from the junction of Río Blanco and Quebrada de la Manga is rich in silt, which surely must be loess, and the lower part shows Stage II+ carbonate buildup (Birkeland, 1984, p. 357-359). This diamicton underlies part of the main terrace surface, which is graded to a surface of aggradation along Río Mendoza, that probably was produced during a major glaciation. The bulbous end of the debris flow, which terminated between La Chacrita and Potrerillos (Fig. 1), protrudes above the surface of the terrace, and its south border rests on the distal edge of an alluvial fan on the south side of Río Blanco. Because of its soil profile characteristics, stratigraphic position, and geomorphic development, I suspect it, also, may have accumulated during a late Quaternary glaciation, when base level along Río Mendoza was higher than it is now, although it surely predates the Late Wisconsinan (Marine Oxygen Isotope Stage 2).

The base of the lowest diamicton exposed in the sequence along the road and stream cuts a short distance west of Potrerillos is only slightly above the modern stream bed. To interpret it as having been left by a debris flow during an interglaciation, when base level was low, would seem reasonable. A complicating factor, though, must be considered. This and other diamicton exposures in this area lie on the downthrown side of a fault (Fig. 1,2), which evidently has been active in late Quaternary time. The diamictons have been offset, but alluvial sediments obscure the trace of the fault across the middle of the valley of Río Blanco.

A bed of volcanic ash caps the diamicton exposed at the base of the steep banks of Río Blanco at Potrerillos (Fig.1, site H) and is in turn overlain by a thick accumulation of coarse gravel (Corte, 1957, Fig. 3; Polanski, 1966, Fig. 2; Wayne and Corte, 1983, Fig. 8). The base of this diamicton lies at the level of the present bed of Río Blanco, but this site, too, is on the downthrown side of the fault that has contributed to preservation of the diamictons downstream from La Chacrita. The volcanic ash dated tentatively as "probably in the 100-200,000 year range" (Glen Izett, letter, 18 Dec. 1980). The zircon microphenocrysts have a low uranium content and a low fission-track density.

Recently, electron microprobe analysis have shown that the chemical composition of glass chards in these volcanic ash lenses is very similar to that of pumice clasts in pyroclastic deposits expelled in a series of closely-spaced eruptions from the Maipo volcano approximately 450,000 years ago (Sterns et al. , 1983). The use of electron microprobe analyses to identify ash beds has become a reliable stratigraphic technique (Smith and Westgate, 1969; Sarna-Wojcicki et al., 1984). The overall similarity of these three ash lenses to the published analyses of the Maipo pumice makes it seem likely that the Maipo volcano was the source of the ash over the diamicton near Potrerillos. A difference in FeO, MgO and CaO, as well as in fission track density in zircons, however, makes it difficult to accept a correlation with the 450 ka pumice from Maipo. In addition, 1987 field studies showed that two volcanic ash lenses separated by alluvial sediments are present in some exposures in the piedmont plain near Mendoza. At this time, then, correlation of the ash bed that overlies the diamicton near Potrerillos with the 450 ka eruption of Maipo can neither be confirmed nor ruled out.

In either event, our suggestion (Wayne and Corte, 1983) that th diamicton at Potrerillos, based largely on the 100-200 ka age of the ash that overlies it, was deposited during a glaciation associated with Oxygen Isotope Stage 6 will need further re-evaluation as more data are accumulated. Should that ash lens be the same as the 450 ka ignimbrites of Maipo, considerably greater time would be available for emplacement of the other diamicton of the Río Blanco valley. Evidence available at this time, however, indicate that it is more likely to be younger.

A greater problem exists with efforts to place the diamictons upstream within a stratigraphic frame. The two exposed along Arroyo Angostura near its junction with Arroyo de los Vallecitos (Fig. 1, site B) seem particularly difficult to fix, since a glacial correlation is not necessarily valid. The soil profile that has developed on the uppermost of the two at this location is a Stage III-IV caliche about 20 to 30 cm thick, which surely formed under conditions more arid than the present at this site. The length of time necessary to develop a carbonate-rich horizon of this character probably was at least several tens of thousands of years. A K horizon of similar characteristics has formed on the surface of the diamicton ridge that stands as an erosional remnant above the

alluvial plain just east of the mountain front, and below
which Río Blanco has entrenched its channel (Fig. 1, site F).
Because of these similarities of soil profile development
and the observations that both remnants are well above
the valley flat (12 to 15 m), they may be isolated parts of
the same deposit. The debris flow that deposited them
probably took place when Río Blanco was graded to a higher
base level than now, perhaps during a glaciation.

Less arid conditions must have existed, though, when the
lower diamicton at site B of Fig. 1 underwent weathering. The
paleosol includes a well developed blocky orange-brown Bt
horizon and shows little accumulation of carbonate (Wayne,
1984, p. 406), although the matrix of the unweathered diamicton
beneath the paleosol contains secondary $CaCO_3$ grains.
These are characteristics of a soil profile that formed under
relatively warm, humid conditions. If so, it would suggest
that the climate of the Cordón del Plata piedmont was
somewhat different during part of Pleistocene time than
it is now.

Because all of these diamictons contain quartz grains that
show, under scanning electron microscope examination,
surface textures associated with glacier transport, this part
of the Andes evidently supported valley glaciers prior to
the last glaciation, although it is unlikely that the
older ice tongues were extensive enough to leave deposits
beyond those that reached about 2600 m during the last
glaciation. Although earlier glaciers may have reached nearly
the same altitudes as those of the Vallecitos (Wisconsinan)
ice in the Río Blanco basin, even slight uplift in this part of the
Andes would have carried them high enough that they now would
be buried beneath the moraines of the last glaciation. Where
younger glaciers overrun the deposits of an older one in alpine
valleys, rarely does a record of the older one remain
identifiable (Gibbons et al., 1984). Recognition of pre-
Vallecitos glaciations in the Cordón del Plata, therefore, is
likely to result only by indirect means such as those presented
here. There can be little doubt, though, that the cirques of
the Cordón del Plata have been excavated several times by
Pleistocene glaciers.

CONCLUSIONS

The diamictons and associated fluvial and airfall sediments
that fill the valley of Río Blanco below the outer moraines of
the last glaciation are here interpreted to be sediments
deposited by large scale debris flows rather than the result
of glacial or neotectonic activity. Some of them, particularly
the older ones, are associated with late Pliocene and/or early
Pleistocene tectonic activity in the Cordón del Plata. Those
that fill the piedmont valley of Río Blanco are more likely
a result of extreme precipitation events and fluctuations in
base level that resulted from outwash deposition and interglacial entrenchment along Río Mendoza. Late Quaternary
faulting has preserved some of the deposits. Correlations are
based primarily on stratigraphic position, surface morphology
and soil profile characteristics.

ACKNOWLEDGEMENTS

I thank Dr. Arturo E. Corte, Instituto Argentino de Nivología
y Glaciología (IANIGLA), for the opportunity to review and
discuss with him in the field the distribution and character
of some of these deposits, as well as the possibilities
regarding their origin, and for having reviewed an early draft
of the manuscript. Dr. Arthur Bloom and Dr. Manfred Strecker,
Cornell University, and Prof. Aleksis Dreimanis, University
of Western Ontario, have made many useful suggestions to help
me clarify the ideas presented here. I also wish to express
my appreciation for the logistical assistance provided by
Francisco von Wuthenau and by the personnel of IANIGLA who
were involved. Naomi Wayne served as a field assistant
throughout the studies. The field investigations on which
this report is based was supported by NSF grants INT 79 20798
and INT 82 2349. Fission track counts were made by Dr. C. Naeser,
(U.S.Geological Survey, Denver) and electronmicroprobe analyses
by Dr. R. Goble (U. of Nebraska-Lincoln).

REFERENCES

Beaty, C. B. 1963. Origin of alluvial fans, White Mountains,
    California and Nevada. **Association of American Geographers
    Annals.** 53:516-535.

Beaty, C. B. 1974. Debris flows, alluvial fans, and a revitalized catastrophism. **Zeitshrift für Geomorphologie, Supplementband.** 21:39-51.

Birkeland, P. W. 1984. **Soils and Geomorphology,** Oxford University Press, New York, 372pp.

Blackwelder, E. 1928. Mudflow as a geologic agent in semiarid mountains. **Geological Society of America, Bulletin.** 39:465-484.

Browning, J. M. 1973. Catastrophic rock slide, Mount Huascarán, north-central Peru, May 31, 1970. **American Association of Petroleum Geologists, Bulletin.** 57:1335-1341.

Brunotte, E. 1983. Zur allochthonen Formung quartärer Fussflächen in Bolsonen W-Argentiniens. **Zeitschrift für Geomorphologie, Supplement band.** 48:203-212.

Clague, J. J., Evans, S. G. and Blown, I. G. 1985. A debris flow triggered by the breaching of a moraine-dammed lake, Klattasine Creek, British Columbia. **Canadian Journal of Earth Science,** 22(10):1492-1502.

Corte, A. E. 1957. Sobre geología glacial Pleistocénica de Mendoza. **Anales del Departamento de Investigaciones Científicas, Universidad Nacional de Cuyo,** 2(2):1-27.

Costa, J. E. and Williams, G. P. 1984. Debris-flow dynamics. **U.S. Geological Survey Open File Report** (videotape).

Dreimanis, A. and Lundqist, J. 1984. What should be called till? in L.K. Konigsson (ed.), **Ten Years of Nordic Till Research. Striae.** 20:5-10.

Eisbacher, G. H. 1982. Mountain torrents and debris flows. **Episodes,** 1982. 4:12-17.

Evenson, E. B., Clinch, J. M. and Stephens, G. C. 1986. The importance of fluvial systems in debris transport at alpine glacial margins. **American Quaternary Association, Program and Abstracts,** 9th Biennial Meeting, p.58-60.

Fisher, R. V. 1971. Features of coarse-grained, high-concentration fluids and their deposits. **Journal of Sedimentary Petrology.** 41:916-927.

Flint, R. F., Sanders, J. E. and Rodgers, J. 1960a. Symmictite: a name for nonsorted terrigenous sedimentary rocks that contain a wide range of particle sizes. **Geological Society of America Bulletin,** 71:507-510.

Flint, R. F., Sanders, J. E. and Rodgers, J. 1960b. Diamictite, a substitute term for symmictite. **Geological Society of America Bulletin,** 71:1809-1810.

Gibbons, A. B., Megeath, J. D. and Pierce, K. L. 1984.
Probability of moraine survival in a succession of glacial
advances. **Geology**, 12:327-330.

Harrington, H. J. 1946. Las corrientes de barros (mud flows)
de "El Volcán", Quebrada de Humahuaca, Jujuy. **Revista
Asociación Geológica Argentina**, 1(2):149-165.

Hsü, K. J. 1975. Catastrophic debris streams (Sturzstroms)
generated by rockfalls. **Geological Society of America
Bulletin**, 86:129-140.

Hsü, K. J. 1978. Albert Heim: Observations on landslides and
relevances to modern interpretations **in** Voight, Barry (eds.),
Rockslides and avalanches, 1: natural phenomena. Elsevier,
Developments in Geotechnical Engineering, 14A:71-93.

Innes, J. L. 1983. Debris flows. **Progress in Physical Geography**,
7:469-501.

Jian, Li, Jianmo, Y., Cheng, B. and Defu, L. 1983. The main
features of the mudflow in Jiang-Jia Ravine. **Zeitschrift
für Geomorphologie N. F.**, 27:325-341.

Keefer, David. 1984. Rock avalanches caused by earthquakes:
source characteristics. **Science**, 223:1288-1290.

Landim, P. M. and Frakes, L. A. 1968. Distinction between tills
and other diamictons based on textural characteristics.
Journal of Sedimentary Petrology, 38:1213-1223.

Lliboutry, L., Arnao, B. M., Pautre, A. and Schneider, B. 1977.
Glaciological problems set by the control of dangerous lakes
in the Cordillera Blanca, Peru. I. Historical failures of
moraine dams, their causes and prevention. **Journal of
Glaciology**, 18:239-254.

Lindsay, J. F. 1968. The development of clast fabric in mudflow. **Journal of Sedimentary Petrology**, 38:1242-1253.

Mills, H. H. 1984. Clast orientation in Mount St. Helens
debris-flow deposits, North Fork Toutle River, Washington.
Journal of Sedimentary Petrology, 54:626-634.

Plafker, G. and Ericksen, G. E. 1978. Nevados Huascarán
avalanches, Peru. **In** Voight, Barry (eds.), **Rockslides and
avalanches, 1: natural phenomena**, Elsevier Sci. Pub.,277-314.

Pierson, T.C. 1986.Flow behavoir of channelized debris flows,
Mount St. Helens, Washington,**in A.D. Abrahams, ed., Hillslope,
Processes, Allen Unwin, Inc,,** Winchester, MA, 269-296.

Polanski, J. 1953. Supuestos englazamientos en la llanura pedemontana de Mendoza. Revista Asociación Geológica Argentina,
8:195-213.

Polanski, J. 1962. Estratigrafía, neotectónica y geomorfología del Pleistoceno pedemontano entre Ríos Diamante y Mendoza. **Rev. Asociación Geológica Argentina,** 17:127-349. Buenos Aires.

Polanski, J. 1965. The maximum glaciation in the Argentine Cordillera. **Geological Society of America, Special Paper** 84, p. 453-472.

Polanski, j. 1966. Flujos rápidos de escombros rocosos en zonas áridas y volcánicas. Editorial Universitaria de Buenos Aires, 67 p.

Polanski, j. 1972. Descripción geológica de la Hoja 24a-b, Cerro Tupungato, Provincia de Mendoza. **Dirección Nacional de Geología y Minería** , Boletín, 124:114 p.

Sarna-Wojcicki, A.M. **et al.** 1984. Chemical analyses, correlations and ages of east-central and southern California. **U.S. Geological Survey, Professional Paper** 1293, 40p.

Sharp, R.P. And L. H. Nobles, 1953. Mudflow of 1941 at Wrightwood, southern California. **Geological Society of America, Bulletin,** 64: 547-560.

Smith, D.W.G. and J.A. Westgate, 1969. Electric probe technique for characterising pyroclastic deposits. **Earth and Planetary Science, Letters** , 5:313-319.

Stern, C.R. **et al.,** 1984. Petrochemistry and age of rhyolitic pyroclastic flows which occur along the drainage valleys of the Río Maipo and Río Cachapoal (Chile) and the Río Yaucha and Río Papagallos (Argentina). **Revista Geológica de Chile** , 23: 39-52.

Wayne, W.J. 1984. The Quaternary succession in the Río Blanco basin, Cordón del Plata, Mendoza province Argentina: an application of multiple relative dating techniques, Mahaney W.C. (ed.) **Quaternary Dating Methods,** Elsevier Sci.Pub.p.389-406.

Wayne, W.J. 1985. Pleistocene stream diversion and static rejuvenation, Cordón del Plata, Mendoza Province, Argentina. **Geological Society of America, Abstracts and Programs.**

Wayne, W.J. and Corte, A.E. 1983. Multiple glaciations of the Cordón del Plata, Mendoza, Argentina. **Palaeogeography, Palaeoclimatology, Palaeoecology,** 42:185-209.

Whalley, W.B. **et al.** ,1983. The magnitude and frequency of large rockslides in Iceland in the Postglacial. **Geogr. Annaler,** 65A:99:110

FEDERICO IGNACIO ISLA
CONICET, Centro de Geología de Costas, Mar del Plata, Argentina

3

Where was the sea-level 30-50,000 years ago?
The Patagonian point of view

ABSTRACT

A mid-Wisconsinan sea-level rise is more or less worldwide accepted by Quaternary scientists. However, there is no agreement about its maximum, reached between 30-40 ka ago (ka= kiloanno= 1,000 years).

In Caleta Valdés (Patagonia), well-preserved articulated shells in living position yielded radiocarbon ages between 29 and 41 ka. These shells were sampled in estuarine deposits between gravelly beach systems. An analysis of the sources of error of these radiocarbon datings, made us definitively accept them in concordance with others sampled from latitudes of 35 to 54 degrees South (provinces of Buenos Aires to Tierra del Fuego).

Similar ages suggesting a higher sea-level were obtained in many places. However, they were denied aducing a method failure, or they were used to establish isostatic uplifts.

From a glaciologic-climatic point of view, a higher sea-level was rejected. However, there is no agreement about climatic conditions for the Middle Wisconsin; moreover, if recently it was suggested that part of Hudson Bay and Western Norway were free of ice. Possibly, similar ice-free conditions had taken place in Western Antarctica and Queen Elizabeth Islands.

In Patagonia, a mid-Wisconsinan higher sea-level is recorded without meaning an excessive isostatic uplift for this trailing-edge coast.

RESUMEN

Un episodio de elevación del nivel del mar durante el Wisconsin medio es más o menos aceptado mundialmente por los científicos cuaternaristas. Sin embargo, no existe aún acuerdo acerca de su máximo nivel alcanzado entre 30 y 40 ka atrás (ka= kiloanno = 1.000 años).

En Caleta Valdés (Patagonia), valvas marinas articuladas y bien preservadas, en posición de vida, proporcionaron edades radiocarbónicas entre 29 y 41 ka. Estas valvas fueron muestreadas en depósitos estuáricos ubicados entre sistemas de playas gravosas. El análisis de las posibles fuentes de error de estas dataciones radiocarbónicas hace factible aceptar definitivamente su concordancia con otras muestras obtenidas en Argentina entre las latitudes 35° a 54° S (provincia de Buenos Aires a Tierra del Fuego).

Edades similares que sugieren un nivel del mar más alto han sido obtenidos en varios lugares. Sin embargo, todas ellas han sido por lo general rechazadas por los investigadores, aduciendo falencias metodológicas o bien, fueron utilizadas para establecer tasas de elevación tectónica.

Desde un punto de vista glaciológico-climático, un nivel del mar más elevado ha sido asimismo rechazado. Sin embargo, no hay acuerdo generalizado acerca de las condiciones climáticas para el Wisconsin Medio; más aún, ha sido sugerido recientemente que parte de la Bahía de Hudson y Noruega Occidental estaban libres de hielo. Posiblemente, condiciones similares libres de hielo habrían tenido lugar en Antártida Occidental y las islas Queen Elizabeth.

En Patagonia, un nivel del mar más elevado durante el Wisconsin Medio ha sido registrado, sin que ello necesariamente signifique una elevación isostática excesiva para esta costa emergente.

INTRODUCTION

Sea-level fluctuations have caused plenty of discussions. The modern trend of relative sea-level change varies globally (Bloom, 1981). The lowest level reached during the last glacial maximum has also generated endless discussion: There are quotes of minimum depths of - 260 m at 15.9 ka ago (Hoshino et al. , 1967, in Emery et al. , 1971) to -60 m at 17 ka (Blackwelder et al., 1979), and they differ from place to place (Curray, 1961; Fray and Ewing, 1963; Emery and Garrison, 1967; Milliman and Emery, 1968; Stearns, 1974; Labeyrie et al. , 1976; Climap Project Members, 1976; Oldale and O'Hara, 1980; Lin, 1982).

The conventional radiocarbon dating limit is more than 55 ka (Bowen, 1978), and 75-80 ka by the enrichment method. A significant controversy began when dates on marine shells in the range of 35-40 ka were obtained from places near or above present sea-level. In many opportunities, the existence of this mid-Wisconsinan "high" sea level has been denied on the basis of radiocarbon contamination with hard waters or humic acids, reworking of old shells, replacement phenomena and recrystallization (Bowen, 1978).

In this paper, we review the evidence of Middle Wisconsinan age raised beaches in the region of Caleta Valdés (Chubut,

Argentina) and we consider the sources of error possibly affecting the shells dates. The eustatic implicancies of these dates have been correlated to other citations around the world. Finally, a revision of paeloclimatic data from 30 to 50 ka ago is presented.

HIGH SEA-LEVELS FROM CALETA VALDES

Caleta Valdés area - as other areas of Patagonian coast -, has gravel ridges deposited as beach berms (Fasano et al. ,1984). These ridges, of more or less the same pattern, consist of five groups. These systems are clearly defined by the elongated ponds between them, suggesting old coastal lagoons or "caletas" (although the lack of freshwater input is recognized). The fifth system - at least - , is Holocene in age and it is distinguishable from those of Pleistocene age by a 10 m step, suggesting a paleocliff (Fig. 1). By means of aerial photographs, it is rather easy to recognize the slightly-curved ponds with a predominant N-S direction. They are surficially in-filled by mud and salt and there is a local relief of more than 10 m to the top of the gravel ridges (Fasano et al, 1984).

At the paelocliff, an outcrop of an old coastal lagoon was observed between systems I and III (Fig. 1). Pelecypod shells in living position. which are very well preserved, were sampled there (Fig. 2) at a height of 7-8 m above present sea level (Fasano et al. , 1984). Radiocarbon dating from these shells yielded ages between 29 and 41 ka (Table I).

CAN RADIOCARBON DATES OF 30-40 ka BE USED ?

The dates from Caleta Valdés suggest a relatively high sea-level of 29-41 ka age. There have been many other suggestions about a mid-Wisconsinan rise in sea level. But very often they have been denied aducing: method failure, shell remobilization, contamination, replacement or recrystallization. Thus, they were assigned to the 120 ka interglacial (Sangamon, Eemian), taking into account a glaciologic-climatic point of view (Mörner, 1971).

The ages obtained from the shells of Caleta Valdés are within the 43 ka limit of the facility at the Instituto Nacional de Geocronología Isotópica (INGEIS) which uses the benzene synthesis technique (Albero et al., 1980).

Shell reworking can be discounted as both valves are still articulated in living position (Fig. 2). As the present water-table is low at Patagonia, contamination with hard waters could be rejected; and higher levels could also be disregarded during the Quaternary if they were not due to a higher sea-level. However, sediment permeability (sand and gravel) around the

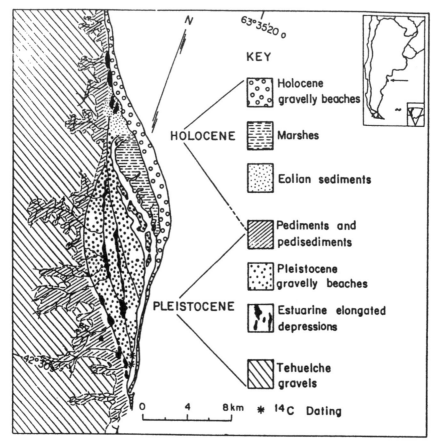

Figure 1. Geomorphic sketch of Caleta Valdés (Chubut, Argentina).
From the 5 gravelly beach systems, the older 4 are Pleistocene.
Between I and III, shells of 34-41 ka radiocarbon age were
samples (Fasano et al . 1984).

shells is significant. On the other hand, $^{13}C/$ ^{12}C ratio did
not indicate any type of contamination.

X-ray diffractograms of the dated shells showed that they
were composed of aragonite and there is no much evidence of
replacement or recrystallization by calcite.

As a preliminary conclusion, it could therefore be said that,
if the datable limit of the Radiocarbon INGEIS Lab is
approximately 43 ka and shells do not exhibit evidence of
remobilization, contamination, replacement or recrystallization
effects, the raised beaches of Caleta Valdés effectively
represent a sea level 7-8 m above present about 30-40 ka B.P.

Figure 2. Articulated pelecypod shell in living position at
Caleta Valdés, and over a gravel deposit.

THE MID-WISCONSINAN SEA-LEVEL RECORD

In 1961 Curray mentioned shoreline deposits 15 m deep in the
shallow shelf off Freeport (Texas), which yielded radiocarbon
ages about 30 ka ago (Table I). He first proposed the idea of
a mid-Wisconsinan interstadial around 30 ka. He also mentioned
Shepard's dates from Hawaii of 24 and 31 ka at heights of
2-4 m above present sea level. He said that although Hawaii
is not stable, the radiocarbon data showed an approximation

of Middle Wisconsin sea-level to present level (Curray, 1961).

Shepard (1963) proposed an eustatic curve which showed sea level at depths of 11-12 m about 26-33 ka ago.

Curray (1965) presented a sea-level curve with a maximum of -10 m at about 29-30 ka which was correlated with an interstadial between 22 and 35 ka.

In the eustatic curve of Milliman and Emery (1968), a minimum is indicated at 15 ka, and a mid-Wisconsinan maximum at 35 ka. In the sense of these authors, this high sea-level does not correspond to an interglaciation but to a fairly cold interstadial at 30 ka (Milliman and Emery, 1968).

However, Mörner (1971) considered that data on sea level 19 m below present and 10 m above present at 34 ka should be considered anomalous and not reliable. Since Mörner's paper, more datings increased the mid-Wisconsinan high sea-level controversy all around the planet (Fig. 3).

1. ARGENTINA AND ANTARCTIC PENINSULA

Bombin (1980) dated at 30 to 35 ka the "Belgranense" estuarine unit which is interbedded within the upper part of the so-called "Pampeano" loessoid sediments. He believed that southeastern South America has been tectonically very stable during Late Pleistocene but some uplift could have occured in the last 30 ka. Thus, it would be possible to explain the discrepancy between this area and others where Middle Wisconsin sea level is below present (Bombin, 1980).

Bayarsky and Codignotto (1982), and Codignotto (1984) recognized ages and positions similar to those of Caleta Valdés at Puerto Lobos and Bahía Bustamante (Chubut), N of Santa Cruz and Tierra del Fuego.

Rabassa (1984) described glaciomarine silts and clays with dropstones and deltaic sands, containing marine shells included in what he called Caleta Santa Marta Drift (James Ross Island, San Martin Peninsula). Shells of Laternula elliptica at 10 m high have been dated at 34 ka. He said that these remains suggest a relatively-warm mid-Wisconsinan episode, in which glaciers had receded and sea level had raised enough so as partly drown Croft Bay, as it is happening now (Rabassa, 1984).

Gonzalez (1984) also mentioned, at the Bahía Blanca area (Argentina), deposits between levels of 11 and 16 m pointing out a higher sea level deposited during the last interstadial of the Wisconsin glaciation (Würm; González , 1984).

2. SOUTH AMERICA

Bittencourt and others (1979) obtained radiocarbon ages of 27 ka from plant remains of marine sands rich in organic matter.

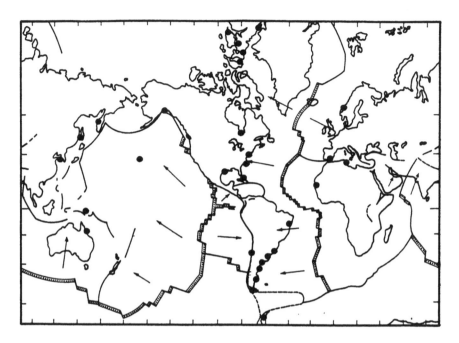

Figure 3. Mid-Wisconsin high-sea-level mentions around the planet.

However, because these remains were found into a permeable material rich in humic acids, Bittencourt et al. (1979) considered them to be possibly much older (Sangamon, 120 ka age) and that they might have been contaminated by modern carbon leached during pretreatment with HONa.

At the Chilean coast, Paskoff (1980) named a Cachagua stadial of 35 ka to a bench located 5-7 m above sea level.

3. NORTH AMERICA

Hopkins (1967) described a marine shell layer of 36 m above sea level at Anchorage, Alaska. It was located between glacial deposits of early and late Wisconsin, and was called the "Woronzofian" transgression. Dated radiometrically by the Th/U method, this layer yielded an age between 33 and 48 ka. Its fossils indicate the influence of arctic waters (Hopkins, 1967). On the other hand, the 7-m high beach deposits of Point Barrow gave radiocarbon ages of 25 ka for organic fibers and peat, and 32 ka for wood fragments (Hopkins, 1967; and Mc Culloch, 1967 in Morner, 1971). However, Morner (1971) mentioned earlier dates of shells from Bootlegger Cove Clay (type section of

Woronzofian transgression) of about 12 ka in height of 7-8m.
So, he corrected the age of this transgression to Late Wisconsin.
Osmond and others (1970) stated that the Outer Merrit Island
barfier (East Coast, USA) would be mid-Wisconsin, with a Th/U
age of 30 ka. However, datings performed by the Ionium and 14 C
method were not consistent, inducing Morner (1971) to think
that these deposits are "... probably of Brorup age" (81 ka).
    Clarke et al. (1972) assigned a marine shell bed, the Salmon
River Sand (Canada), to the last interglaciation because of
its warmer-than-present fauna, despite concordant dates of
38.6 ka (14C) and 44 ka B.P. (Th/U). Grant (1980) accepted the
dates and revised the age to Middle Wisconsinan because of
certain glacial attributes and the enclosing tills. Now, the
Salmon River Sand is recognized as allochthonous, and glacio-
tectonically located in the till sequence (like many of the New
England marine shell beds), without stratigraphic or altitude
significance. It has been re-assigned to the Sangamonian (Grant
and King, 1984). A high-level shell bed on nearby Cape Breton
Island (Grant, 1980) is also proved to be allochthonous based
on foraminiferal assemblages (Guilbault, 1982).
    England (1976) at the Queen Elizabeth Islands (Canada)
mentioned six radiocarbon dates on in situ shells from raised
marine deposits ranging between 28 and 39 ka, some of them as
high as 120 m above sea level (considerable higher than the
inferred postglacial marine limit). He found little reason to
believe these shells to be ice-transported during the last
glaciation (Innuitian ice-sheet), on the basis of their coastal
distribution, uniform age and stratified matrix (England, 1976).

4. AFRICA

Faure and Elouard (1967) obtained four radiocarbon dates from
marine shells of the NW coast of Africa. These dates indicated
that the shells at an elevation of 5 m above MSL are 31-32 ka
old. As they assumed that sea level could have never been higher
than present, they concluded that the area was uplifting at a
rate of 0.5 m/ka, assuming a maximum sea level of -12 m for
31-32 ka B.P. However, Morner (1971) said that so old :
radiocarbon ages should not be trusted. He believed then, that
this high sea-level had to be older (Morner, 1971).
    Einsele et al. (1974) considered the coast of Mauritania
as relatively stable during the Quaternary although they
recognized uplifting trends at Morocco and subsidence at Senegal
coasts. Their measurements of subsidence after the last
transgression (Nouakchottian, Flandrian) was opposite to Faure
and Elouard's (1967) uplifting trends. However, they hesitated
about radiocarbon dates of 30-40 ka of the Inchirian
transgression arguing recrystallization phenomena (Einsele
et al. , 1974).

5. MEDITERRANEAN REGION

At the Chott-el-Djerid (South Tunisia), Richards and Vita Finzi (1982) mentioned marine deposits 35-40 m above MSL. Radiocarbon dates from **Cerastoderma glaucum** and **Ostrea stentina** gave ages of 25.36, 28.95 and 35.14 ka (Table I). These shells have no signs of recrystallization or replacement by calcite. However, as Richards and Vita Finzi believed that sea level was 40 m below present at approximately 30 ka ago (in the sense of Chappel and Veeh, 1978), they concluded that the studied area uplifted 75-80 m. This rate of 2.3-3.3m/ka is conflictive with previous ideas of stability or subsidence (Richards and Vita Finzi, 1982).

Goy and Zazo (1984) dated shells of **S. buboniusat** the East of Almeria (Spain) in 37.72 ka (14 C) and 39 ka (Th/U). They correlated these deposits to the level at +20 m at Faro Roquetas.

6. EUROPE

In seven localities of the Alesund area (Western Norway), Mangerud **et al.** (1981) sampled marine shells in a Till (Rogne Till). Fifteen radiocarbon dates yielded ages between 28 and 38 ka. The shells fragments range from 2 to 4 mm. As they were found in excavations, Mangerud **et al.** reduced a contamination possibility due to groundwater; they also mentioned that equal tills have low permeabilities. They correctly assigned importance to the fact that the ages do not differ significantly in spite of coming from various localities (mainland and 4 islands). Redatings gave concordant ages (Mangerud **et al.**,1981). Amino acid D/L ratios yielded similar results for these shells (Fig. 4), corroborating a mid-Wisconsinan age (Alesund interstadial). However, Th/U datings of two samples yielded ages between 77 and 133 ka, that contradict radiocarbon and amino acid ages (Table 2; Mangerud **et al.**, 1981). Similar ages were obtained at Northern Norway (Mangerud, 1981).

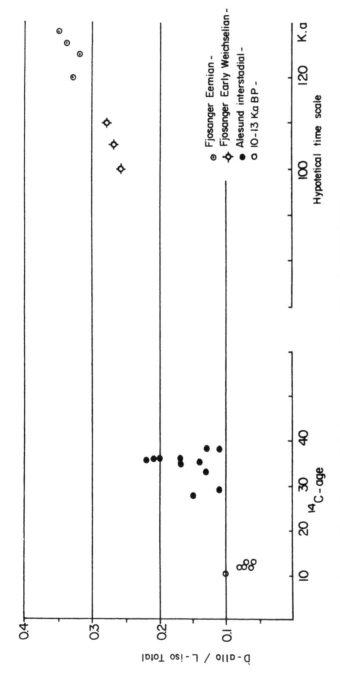

Figure 4. Amino-acid ratios related to radiocarbon data discriminate
the Alesund interstadial samples from those Eemian, Early Weichselian
or Holocene (from Mangerud et al., 1981).

References :
1. (*) Datings used for Fig. 6 fitting, (U) U/Th dates, ( )
   Radiocarbon dates
2.   Isotopic lab and sample number
3.   Altitude, (-5) approximately 5 meters below sea level
4.   Age in ka (1,000 yrs. B.P.), standard errors avoided
5.   Country - Location - Reference

a   Argentina (Pto. Lobos) Bayarsky & Codignotto (1982)
b   Argentina (C. Valdés) Fasano et al. (1982)
c   Argentina (C. Valdés) Codignotto (1983)
d   Argentina (Camarones) Codignotto (1983)
e   Argentina (B. Bustamante) Codignotto (1983)
f   Argentina (S. Cruz Norte) Codignotto (1983)
g   Argentina (T. del Fuego) Codignotto (1983)
h   Argentina (B. Blanca) González (1984)
i   Argentina (La Plata) Bombin(1980)
j   Argentina (Pipinas)Bombin (1980)
k   Argentina (Antarctic Pen.) Rabassa (1984)
l   Argentina (Antarctic Pen.) Clapperton & Sugden (1982)
m   Brazil (Bahía) Bittencourt et al. (1979)
n   Brazil (R. Grande do S.) Bombin (1980)
o   Chile Paskoff (1980)

7. ASIA

Kaplin (1982) recognized several transgressive-regressive
cycles during the Pleistocene in Primorie Coast (USSR), which
he correlated with glacio-eustatic changes. Coastal and
lagoonal deposits with diatoms and warm-water plants were found
in bays at levels similar to present sea level. Radiocarbon
dates yielded ages of more than 50, 40.04 and 41.4 ka. However,
he thought that these ages are too young and that these deposits
were probably from Sangamon age (Riss-Wurm), when sea level was
higher than present (Kaplin, 1982). Younger deposits were
sampled in the same region at depths between 5.8 and 7.3 below
the water-sediment boundary. There, Kaplin (1982) recognized
three horizons:
   1. marine sediments corresponding to deep to shallow-water
facies.
   2. continental sediments with peat lens dated at 29 ka
radiocarbon age and
   3. marine deposits corresponding to climatic conditions
similar to the present ones.
   Kaplin (1982) said that the shallow-water sediments could
be correlated to the Vorontsovo transgression indicating a
warm interglacial. Evidence from the South of Primorie region
would correspond to a maximum level of -16 m for this period
(Kaplin, 1982).

Table I. Late and Mid-Wisconsinan sea-level data from South
America and Amtarctic Peninsula.

1	2	3	4	5
*	Ac 146	15	32.10	a
*	Ac 145	10	30.40	a
*	Ac 144	8	20.30	a
*	Ac 343	7	41.00	b
*	Ac 344	7/8 (7.5)	39.00	b
*	Ac 345	7/8 (7.5)	34.40	b
*	Ac 161	15/25 (20)	38.70	c
*	Ac 164	17/22(20)	38.90	d
*	Ac 165	17/22(20)	32.90	d
*	Ac 168	15/22(18)	32.20	d
*	Ac 150	20/26(24)	31.80	e
*	Ac 151	20/26(24)	30.90	e
*	Ac 154	28/30(29)	36.00	e
*	Ac 155	28/30(29)	37.00	e
	Ac 133		25.80	f
	Ac 490		25.00	f
	Ac 41		39.00	f
*	Ac 134	25/45(30)	27.60	f
	Ac 135		28.70	f
	Ac 209		29.60	f
	Ac 136		31.00	f
*	Ac 207	25/45(35)	28.40	f
*	Ac 208	25/45(35)	27.50	f
*	Ac 485	12/17(15)	31.00	f
*	Ac 454	12/17(15)	32.10	f
	Ac 460		34.40	f
*	Ac 470	19/21(20)	32.60	g
*	Ac 472	19/21(20)	31.90	g
*	Ac 336	14.5	35.50	h
*	Ac 385	12.5	25.70	h
*	GrN 5632	9	34.00	i
*	GrN 5678	10	35.70	i
	GrN		30.00	j
*		10.7	34.12	k
	SRR 1499	94/114	32.16	l
	SRR	94/114	30.60	l
	Bah 563		27.16	m
*	Si 1040	-10	31.35	n
*		5/7 (6)	35.00	o

On the other hand, Kaplin (1982) also found three beds of
lagoonal origin in the Japan Sea. Radiocarbon dating of the
middle lagoonal bed (19-26 m deep at the South and 38-44 at
the North) gave ages of 42, 33.8 and 26.74 ka. He thought that
they must have been correlated to the Vorontsovo transgression
(mid-Wisconsin; Kaplin, 1982).

Kaplin believed as well that the Okhotsk Sea had subsided
while the middle ridge Kamtchatka Peninsula was being uplifted
during the Pleistocene. This is the way he could explain marine
rock benches of mid-Wisconsinan (radiocarbon) age at an
elevation of +20 m above sea level. This was climatologically
supported by pollen assemblages (Kaplin, 1982). But, assuming
that mid-Wisconsinan maximum sea-level was 15-30 m below present,
he concluded that the West coast of Kamtchatka Peninsula had
uplifted approximately 40-50 m. Finally, he mentioned that
these benches at 20-30 m above MSL and a 14 C age of 30-40 ka
are at the same heights that mid-Wisconsinan benches of
Chukotka (Kaplin, 1982).

In Kikai Island (Ryukyu Islands), Konishi et al. (1970)
observed a well-sorted and poorly-cemented calcarenite (Araki
Limestone) at approximately 19-27 m above sea-level. The 230 Th
method yielded ages between 39 and 45 ka. By its content of
Foraminifera and calcareous algae, they assigned a depth of
deposition of -10 m. Assuming a -40 m sea-level for 40 ka ago
(Fairbridge, 1961), Konishi and co-authors (1970) concluded
that Kikai Island has an uplift rate of 1.7-1.8 m/ka. They
said that the limestone age corresponded with the warm interval
at 50 ka postulated by Brouwer and Workom (Konishi et al.   ,
1970).

Liu et al. (1983) obtained cores from the shelf of Huanghai
Sea (China). They recognized a "Lianyangung" transgression
composed of caliches, sands and grey muddy silts with shells
that gave radiocarbon ages greater than 39.5 and 37.4 ka.
Using those data they admitted an age between 25 and 42 ka for
the Lianyangung transgression, correlating it with their
intra-Wurm period (30-35 ka sensu Lin, 1983). Spores and pollen
assemblages (Pinus, Quercus, Artemisia, Chenopodiaceae and
Polipodiaceae) would point towards a cold-temperate climate.
Forams content indicated coastal to shallow marine environments.
They correlated this transgression to the Xian Xian transgression
of the Gulf of Bohai (Zao Songling et al. , 1978 en Liu et al.,
1983) and the Pseudorotalia transgression of the South of China.
Liu and his co-authors (1983) estimated a maximum elevation of
sea level of 16 below the present one for this period.

Table II. Late and Mid-Wisconsinan sea-level data from North America and Hawaii.

1	2	3	4	5
*	L 1380	-35	16.92	a
*	L 1386	-45	20.73	a
*	L 40	-28	29,10	a
*	L 127	-65	22.42	a
*	Gos 1847	-33	25.42	a
*	Gos 1847°	-33	27.65	a
*	Gos 1806	-25	24.20	a
*	1087	-90	19.20	a
*	E 8200	-99/-108(104)	15.18	a
*	Pil 36	-19	21.00	a
*	Gos 1790	-33	17.29	a
*	BB 10a	- 4	33.75	a
*	I 749	10	34.00	a
*	I 1745	0	36.00	a
*	13	-15	28.40	b
*	J 383	-16.5	26.90	c
*	J 526	-16.5	32.50	c
		2/4 (3)	24.00	d
		2/4 (3)	31.40	d
	LJ 205	-1.6	24.14	e
	LJ 206	-1.6	26.64	e
	LJ 253	-4	31.54	e
	LJ 254	-4	31.84	e
	U Gx 1947	2	19.60	f
	U Gx 1948	-1	21.60	f
	U Gx 2838	5	23.22	f
	U	36	33-48	g
*		7	25.00	h
*		7	32.00	h
*	UCLA1146	-34	24.90	i
	GSC 1440	6	38.60	j
	U	6	33-40	j
	GSC 283	20	38.27	j
	GSC 139	42	36.60	k
	GSC 113	105/115	36.80	k
	GSC 65	120	38.60	k
	GSC 149		37.20	k
	GSC 134	85/90	29.80 -	k
	St 4325	100	27.95	k

References (Table II)

1. (*) Datings used for Fig. 6 fitting, (U) U/Th dates, ( )
   radiocarbon dates
2.   Isotopic lab and sample number
3.   Altitude, (-5) approximately 5 meters below sea level
4.   Age in ka (1,000 yrs B.P.), standard errors avoided
5.   Country- Location- Reference

a   East USA Milliman & Emery (1968)
b   East USA (Delaware) Kraft (1971)
c   East USA (Texas) Curray (1961)
d   Hawaii Curray (1961)
e   Hawaii Shepard (1963)
f   Hawaii Stearns (1974)
g   Alaska (Anchorage) Karlstrom (1964 in Morner 1971)
h   Alaska (Pt. Barrow) Hopkins (1967   "   )
i   Mejico   Shepard (1963)
j   Canada   Grant (1980)
k   Canada   Q. Elizabeth Is. England (1976)

8. OCEANIA

Hopley (1971) in Camp Island (North Queensland, Australia)
mentioned a beach-rock platform at approximately +4.5 m of
20.2 ka age (Table I). He explained this age by recrystallization of part of the carbonate. However, in beach ridges and
dunes of the Burdekin delta (North Queensland) other ages of
25.15 and 26.9 ka were obtained. Hopley (1971) explained them
again by leaching and precipitation of modern carbonate.
Similar radiocarbon dates were obtained in the north of
Townsville (Hopley, 1971).
  Stearns (1974) obtained radiocarbon ages from shells of a
beach-rock near Kahuku Point (Oahu, Hawaii) of 19.6, 21.6
and 23.22 (Table I) from what they called Leahi II beach-rock.
This beach-rock is resting on the Kawele soil and underneath
the dunes of Laniloa formation. Dating of snake remains in
such lithifield dunes gave an age of 25.15 ka (HIG-35).
Samples were located between +8 and -1m (Stearns, 1974; Table
4). However, Stearns mentioned that coral fragments from Leahi
II level yielded U ages pointing to a Sangamon interglacial
or Leahi I from Hawaii (115 and 137 ka; Stearns, 1974; and
110 ka, Veeh, 1966). Those samples which gave radiocarbon ages
of 23.2 ka could not be dated by the U method because they
were highly contaminated (Stearns, 1974).

Table III

1	2	3	4	5
*	BETA2653	40	35.14	a
*	BETA2615	40	28.95	a
*	BETA2654	35	25.36	a
*		5	31-32	b
*		-49	25.00	b
*		20	37.72	c
*	U	20	39.00	c
*	T 3938	1/10 (5)	28.49	d
*	T 2306	(5)	28.40	d
*	T 2660	(5)	36.00	d
*	T 2658A	(5)	37.10	d
*	T 2658B	(5)	37.90	d
*	T2658IIA	(5)	37.20	d
*	T2658IIB	(5)	36.50	d
	T 3319		29.00	d
*	T 2659	(0)	30.10	d
*	T2659II	(0)	33.40	d
*	T 2849	(4)	34.70	d
*	T 3120	(-5)	31.30	d
*	T3120II	(-5)	35.10	d
*	T 2657	(5)	35.70	d
*	T2657IIA	(5)	35.10	d
*	T2657IIBC	(5)	35.70	d
	T 2670		34.33	e
	T 2377		40.77	e
*		(0)	40.04	f
*		(0)	41.40	f
		-19/-44	42.00	g
		-19/-44	38.00	g
		-19/-44	26.74	g
*	W 2343	-140	15.74	h
*	W 2216	-64	27.00	h
*	W 2036	-112	15.20	h
*	W 2341	-130	23.26	h
*	H80/19	-16	37.40	i
*	U CK 4	-27	44.00	j
*	U CK 14	-27	38.00	j
*	U CK 15	-27	45.00	j
*	U CK 11	-27	39.00	j
*	U CK 12	-19	42.00	j

Table III (continued)

1	2	3	4	5
*	Gak	4.5	20.20	k
U	LDGO1353C	(62)	35.00	l
U	LDGO1353D	(60)	42.00	l
U	LDGO1347D	(63)	42.00	l
U	LDGO1351E		42.00	l
	ANU 156		29.30	m
	ANU 160		28.50	m
U	ANU 150		46.00	m

Table III. Late and Mid-Wisconsinan sea-level data from Africa, Europe, Asia and Oceania.

References :

1. (*)   Datings used for Fig. 6 fitting, (U) U/Th dates, ( ) radiocarbon dates
2.   Isotopic lab and sample number
3.   Altitude, (-5) approximately 5 meters below sea level
4.   Age in ka (1,000 yrs B.P.), standard errors avoided
5.   Country - Location - Reference

a.   Tunisia Richards& Vita Finzi (1982)
b   Mauritania Faure & Elouard (1967)
c   Spain Toy & Zazo (1984)
d   W. Norway Mangerud et al. (1981)
e   N. Norway Mangerud (1981)
f   USSR (Primorie) Kaplin (1982)
g   Japan Sea Kaplin (1982)
h   East China Emery et al. (1971)
i   Huanghai Sea Liu et al. (1983)
j   Ryukyu Islands Konishi et al. (1970)
k   Australia (N. Queensland) Hopley (1971)
l   New Guinea (Huon Pen.) Bloom et al. (1974)
m   New Guinea (Huon Pen.) Chappell (1974)

DISCUSSION

As it has been pointed above many discussions have been
developed about the occurrence and maximum altitude of mid-
Wisconsinan sea-level rise. Sea levels near or above the
present have been postulated for this period (Hopkins, 1967;
Milliman and Emery, 1968; Paskoff, 1980; Codignotto, 1984;
Fasano et al. , 1984), but their reliability depends on the
datable limit of the radiocarbon method. Other authors accepted
depths of 10 and 20 m (Shepard, 1963; Curray, 1965; Kaplin,
1982; Liu et al. , 1983), or depths between 30 and 40 m
(Fairbridge, 1961; Mörner, 1971; Bloom et al., 1974; Chappell
and Veeh, 1978) for the mid-Wisconsinan maximum.

Other researchers have found sea-level evidences at higher
positions than present but explain them by tectonics (Faure
and Elouard, 1967; Konishi et al.,. 1970; Richards and Vita
Finzi, 1982; Kaplin, 1982).

One of the principal evidences of shorelines of -40/-42 m
for 40 ka came from the Huon Peninsula (New Guinea; Bloom
1974; Chappel and Veeh, 1978). However, the conclusions were
achieved based on the assumption that uniform isostatic uplift
rate had taken place for a period of 140 ka (Chappell, on the
other hand, used a non-linear uplift rate model; 1974);
notwithstanding at the same time, differences of more than
1.5 m/ka of uplift rate are admitted along a distance of 30 km
(2.56 m/ka at Kanzarua and 0.94 m/ka at Kambin). Other
assumptions came from other previous assumptions: shorelines
at 15 and 13 m above present sea level corresponding to 103
and 82 ka (Mesolella et al., 1969) and a uniform uplifting
rate at Barbados (Bloom et al, 1974). A "premature emergence
correction" of the order 3 ka is proposed to obtain the age
of the reefs (Bloom et al, 1974). Chappell and Veeh (1978)
found better correlations of terraces II and IIIb with the
southern margin of North-America Wisconsinan ice-sheet (Fig.5),
than with [18]O analysis of deep-sea cores.

In Patagonia, a neotectonic component should be taken into
consideration (Emery and Garrison, 1967). Nevertheless, it is
hard to believe that neotectonics alone could explain that
shorelines formed at -40 m 40 ka ago, are today higher than
8-10 m above sea level; therefore, indicating an uplift rate
of more than 1.25 m/ka for a trailing edge coast.

High sea levels during glacial cycles could be perfectly
explained by changes on the equipotential surface or geoid
(Mörner, 1976; Bowen, 1978). However, geoidal variations would
probably only explain higher sea levels near ice masses;
elsewhere relative sea level should be lower.

Rapid sea-level rises of the order of 10-30 m in a short
time could also be explained by Antarctic ice surges (Hollin,
1965; 15-20 m sensu Schubert and Yuen, 1982). These events could
be produced by a fast movement of huge ice masses towards the

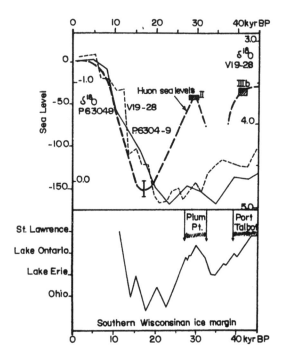

Figure 5. Coral terraces II and IIIb from Huon Peninsula
(New Guinea), related to ^{18}O analysis of deep-sea cores
(V19-28, Panama Basin, and P6304-9, Caribbean Sea), and
southern margin variations of Laurentide ice-sheet (from
Chappell and Veeh, 1978).

sea that could be originated by variations of the orbital
movement of the Earth of episodic volcanic activity (Schubert
and Yuen, 1982). It has been said that these types of
phenomena were recorded in the core of Camp Century (Greenland)
by rapid changes in the relations of $^{18}O/^{16}O$ during the last
interglacial (Dansgaard et al. , 1972). Such cold events at
73, 89.5 and possibly 109 ka have been correlated to high sea
levels (Hollin, 1972). Nevertheless, surges must be linked to
high interglacial sea levels and a collapse of West Antarctica
ice sheet. This means that sea level must be near its present
level to have a record higher than present.

In the present paper, it is preferred to accept mean curves
of sea level during Late Wisconsin of areas relatively stable
than to accept eustatic curves from areas known very unstable.
The exponential curve of the data óf table I fits the equation:

$y = -300 + 86 \ln X$; $r = 0.36$ (Fig. 6)

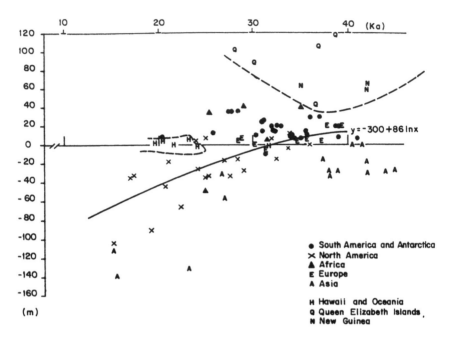

Figure 6. The logarithmic curve for the Làte-Wisconsin sea-level drop. Data from Hawaii, New Guinea (tectonic uplift), and Queen Elizabeth Islands (glacio-isostatic-uplift) were excluded from the fitting.

THE CLIMATIC RECORD

Cold interstadials or colder conditions.
   The Denekamp interstadial of Europe reflects a weather similar to the last glacial epoch (12 ka; Van der Hammen, 1967 in Frerichs, 1968). Van der Hammen and González (1960) arrived to the same conclusions through their palynological studies of northern South America.
   Heusser (1964) said that the Olympia interstadial (Washington, USA) corresponds to a vegetation only 1.5 Celsius degrees colder than present.
   Hopkins (1967) stated that fossils coming from the Woronzofian transgression were indicating cold arctic waters.
   In the sense of Karrow and Terasmae (1970), the Plum Point interstadial is cold with a northern-boreal, if not subarctic, pollinic flora. Dreimanis believed that during that epoch, the ice had reached positions more or less similar to those corresponding to the maximum glaciation (12 ka). Similar conclusions were obtained by Mörner (1969 and 1970, in Mörner, 1971) in southern Scandinavia (Glumslov interstadial).

52

Mörner (1971) analyzed in detail the paleoclimatic evidences
that could be correlated to high sea levels of 24-36 ka. He
said that the Denekamp interstadial of Europe, and Plum Point
of USA - correlated to the Karginsky, Stillfried B, Paudorf,
Glumslov, Olympia and Farmdalian interstadials -, could not
be neither climatically nor glaciologically warm enough to melt
enough ice to rise sea level to its present position (Morner,
1971).

In New Guinea, climatic colder conditions were also detected
for the gap between 28 and 40 ka (Bowler et al.. ,1976; Aharon,
1983). Aharon (1983) proposed a decrease between 2 and 5.7
Celsius degrees with respect to modern mean temperatures
(Fig. 7). Similar results - colder and more humid summers -,
were determined at the Snowy Mountains (southern Australia;
Bowler et al., 1976).

The spores and pollen assemblages that Liu et al. (1983)
found in the Lianyangung transgression would indicate a cold-
temperate climate (Wurm sub-interglacial).

The zones III and IV based on palynological assemblages in
Tagua Tagua (Chile; Heusser, 1983) would indicate a humidity
increase with a decrease in temperature and evaporation respect
to present conditions of the area between 29.8 and more than
43 ka.

WARM INTERSTADIALS OR INTERGLACIALS

A warm interval between 29.5 and 46 ka is reported by Hough
(1950) from a core of the Ross Sea, Antarctica.

The Paudorf soil of Austria (0.3 m thick) represents a short
interstadial somewhat more humid and perhaps warmer than the
previous Fellabrun Complex (Flint and Brandtner, 1961). Although
its base has not been dated, top horizons gave ages of 25.6 and
25.7 ka (Flint and Brandtner, 1961). However, between the
Paudorf soil and the Fellabrun Complex there are loess deposits
of about 31.6 ka which indicate cold conditions ("Middle Wurm
Loess"; Flint and Brandtner, 1961). Later, Van der Hammen
(1967, in Frerichs, 1968) determined the age of the Paudorf Soil
between 27 and 39 ka by stratigraphic correlation.

Similarly, indications of warm climate around 30 ka ago have
been mentioned from southeastern California, Great Lakes-
St. Lawrence region, and Bogota in figure 1 of Flint and
Brandtner's paper (1961). In the Great Lakes-St. Lawrence
region a glacial period is proposed between the Plum Point
interstadial (28 ka) and Port Talbot interstadial (approximately
at 48 ka; Flint and Brandtner, 1961).

The Karginsky "interglacial" of Siberia of mid-Wisconsin age
would have been of a temperature similar to present in the sense
of Kind (1967).

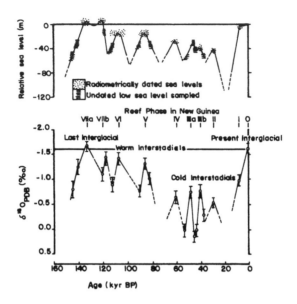

Figure 7.   0 record of the elevated reefs of Sialum
(New Guinea) for the last 160,000 years (from Aharon, 1983).

Frerichs (1968), working with abundance curves of **Globigerina
rubescens** from cores of the northern Indian Ocean (Andaman Sea
and Bay of Bengal), obtained a record of the Pleistocene-
Holocene with a warm interstadial between 22.5 and 42 ka. This
record fits very well with Emiliani's data (1955 and 1970) based
on isotopic temperatures, and Ericson's et al.   data based on
the distribution of **Globorotalia menardii** (1964). However, only
in Emiliani's and Frerichs' curves the late Wisconsinan
interstadial is recorded as a warm interval between 25 and 46
ka (Frerichs, 1968; Fig.8).

The marine beds assigned to the mid-Wisconsin by Kaplin
(1982) and correlated to the Vorontsovo transgression would
correspond to a warm interglacial. In western Kamtchatka
Peninsula there are high sea levels which were paleoclimatically
confirmed (Kaplin, 1982).

DISCUSSION

The mid-Wisconsinan climatic record also seems controversial.
There are evidences reflecting conditions as cold as to the
glacial period (Van der Hammen, 1967 (in Frerichs 1968);
Mörner, 1971); as warm as today (Hough, 1950; Flint and
Brandtner, 1961; Frerichs, 1968; Kaplin, 1982), or a bit colder

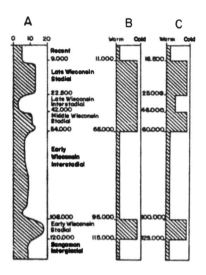

Figure 8. A 25-42 ka-age interstadial was recorded at a) the abundance curve of **Globigerina rubescens** complex (core P-23, Andaman Sea), and c) the isotopic temperature curve of Emiliani (1955). However, this interstadial is not present at the b) abundance curve of **Globorotalia menardii** (Ericson **et al.**, 1964) (from Frerichs, 1968).

(Heusser, 1964; Bowler **et al.**, 1976; Aharon, 1983).

Perhaps, the mid-Wisconsin was a period of climatic instability which needs more detailed studies.

At the Bering Sea, Sancetta and Robinson (1983) studied diatom fluctuations from 3 long cores. They concluded - as Dreimanis and Raukas (1975) had said -, that the Middle Wisconsin was a period of alternating stadials and interstadials conditions. In the Southern Bering Sea, they recognized a limited anual duration of sea ice (perhaps 1 or 2 months), sea-level fluctuations and perhaps of off-shelf transport by rivers. The mid-Wisconsinan flora suggests more variability (seasonality) than today: perhaps more intense winters but summers not different than now (Sancetta and Robinson, 1983).

THE GLACIER RETREAT RECORD

Wilson (1964) suggested that at the Mc Murdo Oasis (Antarctica) the last major glacial retreat was about 50 ka, and that this area was not extensively glaciated during the last northern Hemisphere glaciation.

The evidence of ice margins similar to those of 12 ka age for the Plum Point (Canadian ice-sheet) and Denekamp and Glumslov (Scandinavian ice-sheet) interstadials, together with their short duration, induced Morner (1971) to suppose that climate variations were not so important for a complete deglaciation. Thus, he admitted a maximum sea level of -40 m to -50 m for those ages. He also used conclusions of Flint (1971) assigning an 82 % for the melting effect of the Canadian and Scandinavian (56 and 26 %) ice-sheets into the total volumetric change of the ocean. In the sense of Morner (1971), the melting effect of alpine glaciers around the world and Western Antarctica could not induce to significant changes of his assigned levels of -40 to -50 m. Nevertheless, Morner (1971) explained a possibility of sea level reaching -15 m, if the melting waters would have had discharged so rapidly into the oceans that they would have no time for a subsequent hydroisostatic subsidence.

Using the ages obtained by radiocarbon and amino acid methods, Mangerud et al. (1981) assigned to the Alesund interstadial a middle Weichselian (Middle Wisconsin) age. This ice-free period had arctic conditions - suggested by marine mollucs and benthic foraminifera -, with the Atlantic waters present during the optimal period. Mangerud (1981) said that the faunas of the Alesund and Sandnes interstadials indicate cold environments and the probability of an incomplete deglaciation.

Recently, it has been said - based on previous assumptions as 120 ka age for the Bell Sea marine incursion -, that southern Hudson Bay was also free of ice about 35 ka (Andrews et al. , 1983). This led Andrews and co-authors to take into account the possibility of a "high" sea level during Wisconsin glaciation; which agrees with Fulton's conclusions (1982) that there was no cordilleran ice-sheet between 25 and 45 ka ago (Fig. 9).

Clapperton and Sugden (1982) at the Western Antarctic Peninsula found valves of **Hiatella solida** in a till 94-114 m high. These shells probably represent ice-free conditions before they were dredged up by glaciar ice, as they couldn't exist beneath a thick ice shelf 200 km from the sea. The 35 mm broken shells gave radiocarbon ages of 32.16 (outer) and 30.60 ka (inner fraction; SRR-1499). Amino acid analysis yielded relationships suggesting greater ages and contracdicting a mid-Wisconsinan interstadial. However, Clapperton and Sugden (1982) analyzed two explanations of these ratios based on the probable thermal history of these shells.

Rabassa (1984) also mentioned a mid-Wisconsinan relatively-warm episode in the Antarctic Peninsula, with a glacier retreat and a sea-level rise similar to these days.

DISCUSSION

Morner's glaciologic point of view (1971) against a mid-
Wisconsinan high sea-level was based mainly on:
1. the position of ice during Plum Point and Glumslov-
Denekamp interstadial were comparable to those of 12 ka ago.
2. As the Canadian and Scandinavian ice sheets represent 82
of the "ocean volume changes" (Flint, 1971), even a complete
deglaciation of alpine glaciers and Western Antarctica ice-sheet
was not enough to explain a sea level higher than -40 m.
However, Andrews et al. (1983) admitted the possibility of
a high sea level as the southern Hudson Bay was free of ice
during the Middle Wisconsin. On the other hand, Mangerud
(1981) also considered an incomplete deglaciation at the
Alesund area (Norway). Clapperton and Sugden (1982) and
England (1976) also considered ice-free mid-Wisconsinan
conditions at Western Antarctica and Queen Elizabeth Islands,
respectively.
Flint's (1971) relation between ocean volume changes and
northern Hemisphere ice-sheets during maximum glaciation,
assume an approximately immediate sea-level response to ice
melting in the northern Hemisphere. However, recent studies
have recognized that climatic changes in the southern Hemisphere
possibly preceded those in the northern Hemisphere (Bowler
1976; Lorius et al., 1979). Hays (1978) mentioned that climate
change had taken place 3 ka before in southern Hemisphere than
in northern Hemisphere; this was later corroborated by
Sallinger (1981). Then, Flint's simultaneity assumption would
not be valid at all. Modern glaciologic models need the lags
between northern and southern Hemispheres climatic change, and
then, between climate variation and glacier response.

GENERAL DISCUSSION

The purpose of this discussion is to analyze evidences from
supposed or relatively stable areas in order to accept or
reject the possibility of occurrence of a mid-Wisconsinan
transgression higher than present.
Some authors have rejected these higher sea levels from a
glaciologic-climatic point of view (Morner, 1971), or just
used them to indicate rapid uplifting rates (Konishi et al.,
1970; Richards and Vita Finzi, 1982; Kaplin, 1982).
Today, geoidal variations could explain differences in MSL
of more than 100 m (Mörner, 1976), and there are no objections
to extrapolate these changes to the nearest past. In this
first part of the study of the mid-Wisconsinan transgression
in Argentina, it is believed that to establish the true highest
position reached during this period, it is necessary to work
with MSL curves of the most stable regions of the world.

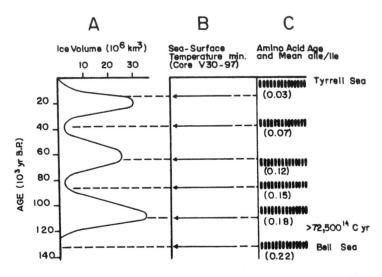

Figure 9. a) Ice-volume variations as a response to orbital
radiation changes at 70°N (Budd and Smith, 1981), compared to
b) sea-surface-temperature minima in the North Atlantic Ocean
(Ruddiman and Mc Intyre, 1981), and c) amino-acid chronology
for Hudson Bay lowlands (from Andrews et al. , 1983).

The effects of Antarctic ice-surges on sea-level changes
should also be taken into account.

Calculations of the volume of water exchange between glaciers
and ocean have assumed an approximately synchronic melting of
ice in both hemispheres. However, a lag of 3 ka between them
has been reported. And, if at the same time, the Canadian and
Scandinavian ice-sheets (and possibly the Innuitian and
Western Antarctica ice-sheets) retreated at least once during
the Wisconsin (Andrews et al. , 1983; Mangerud et al. 1981),
a mid-Wisconsinan higher sea-level could have been recorded
somewhere (Grant, 1984).

Climatic data of the mid-Wisconsin conditions are still very
controversial: colder and warmer temperatures have been
indistinctly reported. Possibly the Middle Wisconsin between
30 and 40 ka was a stage of such a variety of climatic
fluctuations that could not be explained by a single and/or
simple model.

ACKNOWLEDGEMENTS

The author wishes to acknowledge comments provided by E.J.
Schnack, J.L. Fasano and J. Rabassa. Personal communications
with N.A. Mörner, A.L. Bloom, P. Giresse and P.A. Pirazzoli
during the 1984 Intnal. Symp. on Late Quaternary Sea-level

Changes, have also been very useful. But D.R. Grant was the one who gave the greatest impulse and more comments to end the original manuscript.

M.V. Bernasconi, M.Farenga and M. Tomas did the draftings.

CONCLUSION

The question remains without any answer.

If there is an approximate agreement about the sea-level position, climate and ice-extent for 120 ka and 15-13 ka ages, why are there so many doubts about the conditions during the 30-50 ka period ?

REFERENCES

Albero, M.C., Angiolini, F., Balbuena, J.L., Codignotto, J.O., Linares, E. y Weiler, N.E., 1980. Primeras edades Carbono 14 de afloramientos de conchillas de la República Argentina. Rev. AGA, 35, 3, 363-374.

Aharon, P., 1983. 140,000 - yr isotope climatic record from raised coral reefs in New Guinea. Nature, 304, 720-723.

Andrews, J.T., Shilts, W.W. and Miller, G.H., 1983. Multiple deglaciations of the Hudson Bay lowlands, Canada, since deposition of the Missinaibi (Last interglacial?) Formation. Quater. Res., 19, 18-37.

Bayarsky, A. and Codignotto, J.O., 1982. Pleistoceno-Holoceno marino en Puerto Lobos, Chubut. Rev. AGA, 37, 1, 91-99.

Bittencourt, A.C.S.P., Martin, L., Boas, G.S.V. and Flexor, J.M., 1979. Quaternary marine formations of the coast of the State of Bahía, Brazil. In Suguio, K., Fairchild, T.R., Martin, L. and Flexor, J.M. (eds.) Proc. 1978 Intnal. Symp. on Coastal Evolution in the Quater. IGCP, Sao Paulo, 232-253.

Blackwelder, B.W. , Pilkey, O.H. and Howard, J.D., 1979. Late Wisconsin sea levels on the southeast U.S. Atlantic shelf based on in-place shoreline indicators. Science, 204, 618-620.

Bloom, A.L., 1981. 1980 annual report on scientific progress; sea-level movements during the last deglacial hemicycle (about 15,000 yrs). IGCP, Project 61, 23.

Bloom, A.L., Broecker, W.S., Chappell, J.M.A., Matthews, R.K. and Mesolella, K.J., 1974. Quaternary sea level fluctuations on a tectonic coast: new ^{230}Th/^{234}U dates from Huon Peninsula, New Guinea. Quater. Res., 4, 185-205.

Bombin, M., 1980. Evidence for a higher sea level stand in southeastern South America during Mid-Wisconsinan times.6 th. biennial Meet., Abs. and Prog., Am. Quater. Assoc., Orono, Maine, 37.

Bowen, D.Q., 1978. Quaternary geology: a stratigraphic framework for multidisciplinary work. Pergamon Press, London, 221.

Bowler, J.M., Hope, G.S., Jennings, J.N., Singh, G. and Walker, D., 1976. Late Quaternary climates of Australia and New Guinea. Quater. Res., 6, 359-394.

Budd, W.F., and Smith, I.N., 1981. The growth and retreat of ice sheets in response to orbital radiation changes. In Allison, I. (ed.) Sea level, ice and climatic change. IAHS, Publi. 131, 369-409.

Chappell, J., 1974. Geology of coral terraces, Huon Peninsula, New Guinea: a study of Quaternary tectonic movements and sea-level changes. Geol. Soc. Am., Bull. 85, 553-570.

Chappell, J. and Veeh, H.H., 1978. ^{230}Th/ ^{234}U age support of an interstadial sea level of -40 m at 30,000 yr B.P. Nature, 276, 602-603.

Clapperton, Ch. M. and Sugden, D.E., 1982. Late Quaternary glacial history of George VI Sound area, West Antarctica. Quater. Res., 18, 243-267.

Climap Project Members, 1976. The surface of the ice-age Earth. Science, 191, 4232, 1131-1137.

Codignotto, J.O., 1984. Depósitos elevados y/o de acreción Pleistoceno-Holoceno en la costa fueguino-patagónica. Actas, Simp. Osc. del Nivel del Mar durante el Ultimo Hemiciclo Deglacial en la Argentina. CAPICG (IGCP 61), UNMDP, Mar del Plata, 12-26.

Curray, J.R., 1961. Late Quaternary sea level: a discussion. Geol. Soc. Am., Bull., 72, 1707-1712.

Curray, J.R., 1965. Late Quaternary history, continental shelves of the United States. In Wright, H.E. and Frey, D.G. (eds.) The Quaternary of the United States. Princeton Univ. press, 723-735.

Dansgaard, D.W., Johnsen, S.J., Clausen, M.B. and Langway, C.C., 1972. Speculations about the next glaciation. Quater. Res., 2, 376-398.

Dreimanis, A. and Raukas, A., 1975. Did middle Wisconsin, middle Weichselian, and their equivalents represent an interglacial or an interglacial complex in the Northern Hemisphere ? Quater. Studies, Royal Soc. New Zealand, Bull., 13, 109-120.

Einsele, G., Herm, D. and Schwarz, H.U., 1974. Sea level fluctuation during the past 6,000 yr at the coast of Mauritania. Quater. Res., 4, 282-289.

Emery, K.O. and Garrison, L.E., 1967. Sea levels, 7,000 to 20,000 years ago. Science, 157, 684-687.

Emery, K.O., Niino, H. and Sullivan, B., 1971. Post-Pleistocene levels of the East China Sea. In Turekian, (ed.) Late Cenoxoic glacial ages. Yale Univ. Press, New Haven, 381-390.

Emiliani, C., 1955. Pleistocene temperatures. Jour. Geol., 63, 6, 538-578.

Emiliani, C., 1970. Pleistocene paleotemperatures. Science, 168, 822-825.

England, J., 1976. Late Quaternary glaciation of the Eastern Queen Elizabeth Islands, NWT, Canada: alternative models. quater. Res., 6, 185-202.

Ericson, D.B., Ewing, M. and Wollin, G., 1964. The Pleistocene epoch in deep sea sediments. Science, 146, 723-732.

Fairbridge, R., 1961. Eustatic changes in sea level. Phys. Chem. Earth, 4, 99-185.

Fasano, J.L., Isla, F.I. and Schnack. E.J., 1984. Un análisis comparativo sobre la evolución de ambientes litorales durante el Pleistoceno tardío-Holoceno: Laguna Mar Chiquita (Buenos Aires) - Caleta Valdés (Chubut). Actas, Simp. Osc. del Nivel del Mar durante el Ultimo Hemiciclo Deglacial en la Argentina. CAPICG (IGCP 61), UNMDP, Mar del Plata, 27-47.

Faure, H. and Elouard, P., 1967. Schema des variations du niveau de l'océan Atlantique sur le côté de l'Ouest de l'Afrique depuis 40.000 ans. Comp. Rend., Acad. Sci. Paris, 265, Serie D, 784-787.

Flint, R.F., 1971. Glacial Quaternary geology. J. Wiley, 892.

Flint, R.F. and Brandtner, F., 1961. Climatic changes since the last interglacial. Am. Jour. Sci., 259, 321-328.

Fray, C. and Ewing, M., 1963. Pleistocene sedimentation and fauna of the Argentine shelf. I. Wisconsin sea level as indicated in Argentine continental shelf sediments. Proc. Acad. Nat. Sci. Phila., 115, 6, 113-126.

Frerichs, W.E., 1968. Pleistocene-Recent boundary and Wisconsin glacial biostratigraphy in the Northern Indian Ocean. Science, 159, 1456-1458.

Fulton, R.J., 1982. Review of Quaternary stratigraphy, Canadian Cordillera. Geol. Assoc. Can., Abs., 7, 51.

Gonzalez, M.A., 1984. Depósitos marinos del Pleistoceno superior en Bahía Blanca, provincia de Buenos Aires. IX Congr. Geol. Arg., Actas, III, AGA, Bariloche, 538-555.

Goy, J.L. and Zazo, C., 1984. Evolución geomorfológica costera desde el último interglacial a la actualidad en el litoral occidental de Almería (España). Intnal. Symp. on Late Quater. Sea-level Changes and Coastal Evolution, Abs., IGCP-INQUA, Mar del Plata, 52-55.

Grant, D.R., 1980. Quaternary sea-level change in Atlantic Canada as an indication of crustal delevelling. In Mörner, N.A. (ed.) Earth rheology, isostasy and eustasy, Wiley, 201-214.

Grant, D.R., 1984. Conference report. Shore-level change in Patagonia. Litoralia, Newsletter 10, Dec. 84, 3-4.

Grant, D.R. And King L.H., 1984. A stratigraphic framework for the Quaternary history of the Atlantic provinces. In Fulton, R.J. (ed.) Quaternary stratigraphy of Canada. Can. Contr. IGCP Project 24, Geol. Surv. Can., Paper 84-10, 173-191.

Guibault, J.P., 1982. The pre-late Wisconsinan foraminiferal assemblages of St. Lawrence Bay, Cape Breton Island, Nova Scotia. Current Res., part C, Geol. Surv. Can., Paper 82-1C, 39-43.

Hays, J.D., 1978. A review of the Late Quaternary climatic history of Antarctic Seas. In Van Zinderen Bakker, E.M. (ed.) Antarctic glacial history and world paleoenvironments. Xth. Inqua Congr., Rotterdam, 557.

Heusser, C.J., 1964. Palynology of four bogs sections from the western Olympic Peninsula, Washington. Ecology, 45, 23-40.

Heusser, C.J., 1983. Quaternary pollen record from Laguna Tagua Tagua, Chile. Science, 219, 1429-1432.

Hollin, J.T., 1965. Wilson's theory of ice ages. Nature, 208, 12-16.

Hollin, J.T., 1972. Interglacial climates and Antarctic ice surges. Quater. Res., 2, 401-408.

Hopkins, D.M., 1967. Quaternary marine transgressions in Alaska. In Hopkins, D.M. (ed.) The Bering Land bridge. Stanford Univ. Press, Stanford, Ca., 47-90.

Hopley, D., 1971. Sea level and environment changes in the Late Pleistocene and Holocene in North Queensland, Australia. Quaternaria, 15, INQUA, Roma, 267-273.

Hough, J.L., 1950. Pleistocene lithology of Antarctic ocean-bottom sediments. Jour. Geol., 58, 3, 254-260.

Kaplin, P.A., 1982. Sea level changes in the far eastern seas of the USSR in the Pleistocene and Holocene. In Colquhoun, .D.J. (ed.) Holocene sea level fluctuations: magnitude and causes. IGCP-INQUA, Columbia, SC, 96-103.

Karrow, P.F. and Terasmae, J., 1970. Pollen-bearing sediments of the St. Davids buried valley fill at the Whirlpool, Niagara River gorge, Ontario. Can Jour. Earth Sci., 7, 2, 539-541.

Kind, N.V., 1967. Radiocarbon chronology in Siberia. In Hopkins, D.M. (ed.) The Bering Land bridge, 172-192.

Konishi, K., Schlanger, S.O. and Omuba, A., 1970. Neotectonic rates in the Central Ryukyu Islands derived from ^{230}Th coral ages. Mar. Geol., 9, 4, 225-240.

Labeyrie, J., Lalou, C., Monaco, A. and Thommeret, J., 1976. Chronologie des niveaux eustatiques sur le coté du Roussillon de -33.000 ans à nos jours. Comp. Rend., Acad. Sci. Paris, 282, serie D, 349-352.

Lin, D.K., 1982. Variation of Holocene sea level on the coast of Fujian, P.D. de China. In Colquhoun, D.J. (ed.) Holocene sea level fluctuations: magnitude and causes. IGCP-INQUA Columbia, SC, 106-117.

Lin D.K., 1983. Progress of sea level variation research in China during Late Quaternary. Intnal. Symp. on Coastal Evolution in the Holocene, Abs., JSPS, Tokyo, 69-77.

Liu, M., Wu, S., Wang, Y. and Gao, J., 1983. Late Quaternary sea-level changes in Huanghai Sea. Intnal. Symp. on Coastal Evolution in the Holocene, Abs., JSPS, Tokyo, 78-85.

Lorius, C., Merlivat, L., Jouzel, J. and Pourchet, M., 1979. A 30,000 -yr isotope climatic record from Antarctic ice. Nature, 280, 644-648.

Mangerud, J., 1981. The early and mid Weichselian in Norway: a review. Boreas, 10, 381-393.

Mangerud, J., Gulliksen, S., Larsen, E., Longva, O., Miller, G.H. , Sejrup, H.P. and Sonstegaard, E., 1981. A middle Weichselian ice-free period in Western Norway: the Alesund interstadial. Boreas, 10, 447-462.

Mesolella, K.J., Matthews, R.K., Broecker, W.S. and Thurber D.L., 1969. The astronomic theroy of climatic change: Barbados data. Jour. Geol, 77, 250-274.

Milliman, J.D. and Emery, K.O., 1968. Sea levels during the past 35,000 years. Science, 162, 1121-1123.

Mörner, N.A., 1971. The position of ocean level during the interstadial at about 30,000 B.P.: a discussion from a climatic-glaciologic point of view. Can. Jour. Earth Sci., 8, 132-143.

Mörner, N.A., 1976. Eustasy and geoid changes. Jour. Geol. 84, 2, 123-151.

Oldale, R.N. and O'Hara, C.J., 1980. New radiocarbon dates from the inner continental shelf off southeastern Massachusetts and a local sea-level-rise curve for the past 12,000 yrs. Geology, 8, 102-106.

Osmond, J.K., May, J.P. and Tanner, W.F., 1970. Age of the Cape Kennedy barrier-and-lagoon complex. Jour. Geophys. Res., 75, 469-479.

Paskoff, R.P., 1980. Late Cenozoic crustal movements and sea level variations in the coastal area of northern Chile. In Morner, N.A. (ed.) Earth rheology, isostasy and eustasy. J. Wiley, 487-495.

Rabassa, J., 1984. Galciomarine deposits and isostatic recovery in Northern James Ross Island, Antarctic Peninsula. Intnal. Symp. on Late Quaternary Sea-level Changes and Coastal Evolution, Abs., IGCP-INQUA, Mar del Plata, 86.

Richards, G.W. and Vita Finzi, C., 1982. Marine deposits 35,000-25,000 years old in the Chott el Djerid, southern Tunisia. Nature, 295, 54-55.

Ruddiman, W.F. and Mc Intyre, A., 1981. Oceanic mechanisms for amplification of the 23,000- year ice-volume cycle. Science, 212, 4495, 617-627.

Sallinger, M.J., 1981. Paleoclimates north and south. Nature, 291, 106-107.

Schubert, G., and Yuen, D.A., 1982. Initiation of ice ages by creep instability and surging of the East Antarctic ice sheet. Nature, 296, 127-130.

Sancetta, C. and Robinson, S.W., 1983. Diatom evidence on Wisconsin and Holocene events in the Bering Sea. Quater. Res. , 20, 232-245.

Shephard, F.P., 1963. Thirsty-five thousand years of sea level. In Clements, T., Stevenson, R.E. and Halmos, D.M. (eds) Essays of southern California in honor of K.O. Emery. Los Angeles, Univ. of S. California Press, 1-10.

Stearns, H.T. , 1974. Submerged shorelines and shelves in the Hawaii Islands and a revision of some eustatic emerged shorelines. Geol. Soc. Am., Bull,, 85, 795-804.

Van der Hammen, Th. and Gonzalez, E., 1960. Upper Pleistocene and Holocene climate vegetation of the "sabana de Bogotá" (Colombia, South America). Leidse Geol. Mededeel., 25, 261-315.

Selected papers of the final meeting of the International Geological Correlation Program (IGCP) Project 201, Quaternary of South America

Edited by
MONICA SALEMME
CONICET and Museo de La Plata, La Plata, Argentina
JORGE RABASSA
CADIC-CONICET, Ushuaia, Tierra del Fuego, Argentina

IGCP - 201

Computer editing by
María Elena Ardanch and Jorge Ontivero
CADIC-CONICET, Ushuaia, Tierra del Fuego, Argentina

JOSE A. BONINSEGNA
Laboratory of Dendrochronology, Centro Regional de Investigaciones Cientificas y Técnicas, Mendoza, Argentina

4

Santiago de Chile winter rainfall since 1220 as being reconstructed by tree rings

ABSTRACT

The reconstruction of the winter rainfall series of Santiago de Chile back to the year 1220 using the tree-ring chronology of El Asiento is discussed in this paper.

The use of new statistical techniques in the reconstruction of the chronology, in the control of the homogeneity of the sampled site and a careful verification of the results have allowed to improve the percentage of the explained variance between the reconstructed series and the observed one.

This control is extended to the domain of the frequency by the use of crosspectra, coherence and gain analysis.

Some features that are shown in the reconstructed series are discussed.

RESUMEN

Se discute en este trabajo la reconstrucción de las series de precipitación invernal de Santiago de Chile, en forma retrotraída al año 1220 de nuestra era, utilizando la cronología de anillos de árboles de la localidad de El Asiento.

El uso de nuevas técnicas estadísticas en la reconstrucción de la cronología, en el control de la homogeneidad de la localidad muestreada y una verificación cuidadosa de los resultados, ha permitido mejorar el porcentaje de la varianza explicada entre las series reconstruidas y la observada.

Este control se extiende al dominio de la frecuencia mediante el uso de análisis de espectros cruzados, de coherencia y de ganancia.

Se discuten algunos rasgos que aparecen en las series reconstruidas.

INTRODUCTION

The rainfall series of Santiago de Chile is one of the longest records in the area, and it was used in several papers related to the past climate, the distribution and fluctuation of the precipitation in Central Chile (Lamb, 1972; Benitez, 1973; Ereño and Hoffman, 1978; La Marche, 1975; Quinn and Neil, 1982; Vargas and Compagnucci, 1985).

Taulis (1938) and Lamb (1972) have attempted to reconstruct this series using historical information.

Near Santiago de Chile tree-ring chronologies were made by La Marche et al. (1978). Tree-ring widths are well-known as indicator of past climate, in fact they are one of the few proxi- climate variable capable of giving accurate information about the year-to-year scale.

One preliminary reconstruction of the rainfall series of Santiago de Chile was made by La Marche (1975), but only the 19% of the variance was explained by the model.

In this paper, it is made an attempt to improve that reconstruction using modern techniques in the development of the tree- ring chronologies and the statistical verification of the results.

METHODS

1 DATA SOURCE

The monthly rainfall data of Santiago de Chile was obtained from CORFO (1971) and the monthly mean temperature from Jones et al. (1986).

Rainfall record is considered homogeneous according to Benitez (1973) and Vargas and Compagnucci (1985).

Two chronologies of tree-ring width -El Asiento and San Gabriel- were studied as possible sources of data to reconstruct the precipitation (La Marche et al., 1978). Data about tree rings were kindly supplied by Dr. V. La Marche and Dr. R.

Holmes of the Tree Ring Laboratory, University of
Arizona.

The location of the Meteorological Station and
the chronology sites are shown in Figure 1.

In a first step, the crossdate of the tree-
ring data from both El Asiento and San Gabriel was
checked up using the Cofecha program (Holmes,
1983). If one core shows the presence of flags, a
poor correlation with the master series or a record
length lower than 200 years, this particular core
will be rejected from the original chronology.

In the case of El Asiento, from the 65 cores
that form the published chronology, 14 cores were
rejected. In the same way, from 67 cores in the
original set, 11 were separated in San Gabriel. The
improvement in the crossdate, measured as the
increase in the index of mean correlation between
the cores and the master series was of 2.3% in El
Asiento and of 2.4% in San Gabriel.

Then, the new chronologies were built using
the Arstan program (Cook and Holmes, 1984). This
program provides several features that are new in
dendrochronology. The techniques of cubic spline
curve fitting, two stage removal of the growth
trend, biweight robust estimation of the mean value
function, autoregressive modeling and returning the
common pooled persistence to the residuals are used
to improve the common signal contained in a set of
tree rings series.

2 RESPONSE FUNCTION

In order to establish a significant link between
the variations in the tree rings and the climate,
it was used the response function, a multiple
linear regression equation in which the ring width
is the dependent variable and monthly temperature
and precipitation values are the independent ones.

A matrix was built using a set of 16 monthly
temperatures, 16 monthly precipitation data and the
chronology lagged in 1 to 3 which permit to solve
the equation (1)

(1) $Y_{(t)} = A + B_1X_1 + ..$

$+B_nX_n+....+B_aY_{(t-1)}+$

$+B_bY_{(t-2)}+B_cY_{(t-3)}+E$

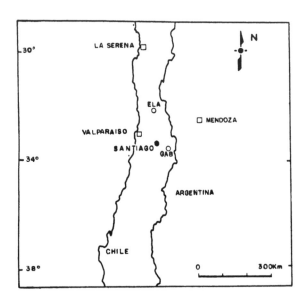

Figure 1. Map of chronology sites. ELA: El Asiento.
GAB: San Gabriel. Meteorological Station at
Santigo.

where $Y_{(t)}$ = tree-ring chronology data for
                      the year t

       X_n = monthly   temperature   or
                      precipitation value

       B_n = regression coefficient

        A   = constant term

        E   = error

     To avoid colinearity problems, principal
components were obtained and the eigenvectors
amplitudes   were   used   replacing   the   original
variables. Because of the transformation of the
data,   the   variables   are   now   orthogonals   and
consequently   fulfill   the   major   assumption   of

statistical independence of the data. The number of variables are reduced selecting those eigenvectors which explain more than the 1% of the total variance.

However, the use of the amplitudes instead of the original variables somewhat obscures the physical relationship of the climate effect upon ring width, so after the equation is solved, a new transformation was carried out in order to express the regression coefficients in terms of the original variables rather than the eingenvector amplitudes (Fritts et al., 1971).

The response functions were calculated using the Response program (Lough, J. 1983).

3 TRANSFER FUNCTION

The transfer function was obtained solving the multivariate equation (2) by means of the Multiple Regression Program of the Statistical Package for the Social Sciences (SPSS).

$$(2) \quad Y_t = A + B_1 X_{(1,t)} + ..$$

$$B_2 X_{(2, t)} + B_1 X_{(1, t)} + E$$

where:

Y_t = predictable variable

X_1 = predictor variables (ring width chronologies)

B_1 = multiple regression coefficient for the i variable

A = constant

E = error

4 CALIBRATION AND VERIFICATION

The calibration of the transfer function was done by splitting the series in two periods and using one to calibrate and the other to verify the fitting. In a second step, the portion used to calibrate was employed to verify and vice versa.

Statistical verification was made testing the goodness of the fitting with the Verify program (Gordon and LeDuc, 1981).

RESULTS

1 RESPONSE FUNCTIONS

The response functions of both sites are shown in Figure 2. San Gabriel´s chronology has a very weak climatic signal that only explains the 24% of the variance for the period between 1922-1970. Instead, El Asiento´s chronology has a signal that permits to explain the 44% of the variance for the same period.

Besides, the structure of the correlation and response functions are quite different. El Asiento shows a strong signal to the winter precipitation while San Gabriel has a more significant response to summer rainfall and a very high influence of the prior growth. Neither of both chronologies presented significant response to temperature parameters.

The conclusion was that only El Asiento has a signal to the precipitation of Santiago which is useful to reconstruct the series.

2 IMPROVEMENT OF THE CLIMATIC SIGNAL

Following the works of Peter et al. (1981), and Pittock and Cropper (1982), the next step was to search alternative ways to improve the climatic signal in El Asiento´s chronology.

In order to eliminate undesirable effects of differences in the autocorrelation structure of each core, the pre-whitening indices provided by ARSTAN program (Cook, 1986) were used.

The principal components of the matrix formed by the indices of each core in the rows and the years for the period 1870-1960 (common period) in the columns were obtained. The frequency distribution of the loading factors suggests the presence of some heterogeneity in the samples; therefore, one subset was made with those cores which have loading factors higher than 0.15. This subset is labelled as PCLF (Table 1) (Peter et al., 1981).

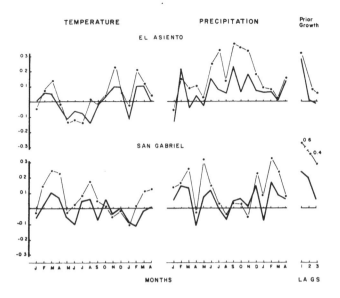

Figure 2. Response function (solid line) and correlation function (dotted line) of El Asiento and San Gabriel.

The cores that have a simple correlation coefficient against the chronology higher than the mean correlation index of all cores (r = 0.63) were put in another set. The subset is labelled

The response function of each core was made using as climatic parameters the log10 of the winter rainfall, the log10 of summer precipitation, and the winter and summer mean temperatures. By this way, those cores that show a relationship with climate, measured as the adjusted squared regression coefficient higher than the mean R2 value (R2 = 0.20) were separated in one subset labelled as RSQR.

Another subset was made with the cores that have a higher mean correlation coefficient against the log10 of winter rainfall (r = 0.435). The subset is labelled HPPT.

Finally, one subset was made with all the cores rejected by the criteria mentioned above (Table 1). It seems that these cores are pathological in some ways and they probably

Table 1. Selection of new subsets: a) adjusted R2 multiple correlation coefficient between cores and the summer and winter temperature and log10 of the rainfall climatic variables in summer and winter (RSQR chronology); b) correlation coefficient of each core against the main chronology (HRCH chronology); c) correlation coefficient of each core against log10 winter rainfall (HPPT chronology); d) 1st principal component of loading factors for the common period (PCLF chronology). The asterisc (*) indicates the rejected cores.

Core ID	a	b	c	d
ELA021	0.2190	0.7100	0.3900*	0.1930
ELA022	0.2260	0.6900	0.4000*	0.1880
ELA031	0.4290	0.8000	0.6600	0.2126
ELA033	0.2270	0.6200*	0.4400	0.1687
ELA041	0.3500	0.7900	0.6000	0.2122
ELA042	0.3200	0.7500	0.5700	0.1999
ELA043	0.1660*	0.7500	0.4300*	0.2033
ELA082	0.0200*	0.4500*	0.1800*	0.1269*
ELA101	0.1950*	0.6100*	0.3600*	0.1686
ELA102	0.2390	0.6000*	0.5300	0.1675
ELA111	0.2300	0.6100*	0.5300	0.1583
ELA131	0.1780*	0.6400	0.4700	0.1795
ELA161	0.0880*	0.6200*	0.2500	0.1690
ELA171	0.2720	0.7000	0.5600	0.1822
ELA182	0.2860	0.6900	0.5500	0.1749
ELA191	0.2070	0.7200	0.4700	0.1949
ELA241	0.2360	0.6900	0.4800	0.1848
ELA243	0.1590*	0.7100	0.4600	0.1843
ELA252	0.0700*	0.5800*	0.3400*	0.1568
ELA253	0.0800*	0.5200*	0.4100*	0.1435*
ELA261	0.0700*	0.4800*	0.2500*	0.1248*
ELA271	0.1500*	0.5200*	0.4200*	0.1423*
ELA272	0.1700*	0.6000*	0.4200*	0.1683
ELA301	0.1300*	0.7800	0.4300*	0.2037
ELA302	0.2550	0.7500	0.4300*	0.1956
ELA321	0.1000*	0.5900*	0.3100*	0.1572
ELA331	0.2100	0.5800*	0.4600	0.1569
ELA332	0.2400	0.6800	0.4900	0.1716
ELA342	0.2170	0.5200*	0.4300*	0.1421*
ELA343	0.2400	0.7000	0.4300*	0.1854
ELA362	0.2980	0.7500	0.5400	0.1985
ELA372	0.2100	0.6400	0.5000	0.1798

introduce some kind of noise into the chronology. The subset is labelled as BADC.

To test this hypothesis, a new chronology was built with each subset, and its characteristics, response function and power to reconstruct the winter rainfall series were established.

The reconstruction was performed using the equation (3)

(3) $Y_t = A + 10 \exp (bX_t)$

where Y_t is the estimated precipitation,

        X_t is the tree-ring

data, b is the regression coefficient and A is a constant term. The results are shown in Table 2.

The chronology built with the samples which have the highest correlation against the log10 of winter rainfall (HPPC) gives the higher explained variance, so it was selected to reconstruct the series.

3 CALIBRATION AND VERIFICATION

Calibration is the formalization of the climate-growth relationship in an equation, linking the climate and tree rings chronologies and allowing the reconstruction of the former from the latter.

Statistical verification is the testing of the reconstructed data by statistical comparison with independent climate data not used in calibration.

The tree-ring data were calibrated against 60 yr of the Santiago´s winter rainfall record (1870-1929) and verified against a set of 31 independent data (1930-1960). The calibration and verification periods were then reversed to check the stability of the results; therefore, in this case, the tree-ring data were calibrated against the rainfall series in the years 1901-1960 and verified for the years 1870-1900.

In a second exercise, the calibration was made using the entire winter rainfall record (1870-1960) and the reconstruction based on this calibration was tested on the first 31 years and on the last 31 years of the record.

A series of verification tests were performed

Table 2. Statistical of chronologies made with the
new subsets.

Chronology	RSQR	HRCH	HPPT	PCLF	BADC
Mean	1.000	1.000	1.000	1.000	1.000
Median	1.007	1.003	1.003	1.007	0.997
Mean Sens.	0.242	0.240	0.238	0.197	0.209
Std. Dev.	0.247	0.249	0.239	0.185	0.188
Skewness	1.203	1.218	1.198	-0.629	-0.131
Kurtosis	24.360	22.820	24.940	4.310	7.033
Autocorrelation					
1st order	0.070	0.089	0.007	0.035	0.008
2nd order	-0.050	-0.050	-0.049	-0.020	0.009
3th order	0.019	0.013	0.013	0.041	0.030
Mean correlation					
radii	0.453	0.495	0.461	0.446	0.300
trees (Y)	0.442	0.481	0.454	0.438	0.297
within trees	0.731	0.793	0.689	0.720	0.330
Signal/noise ratio	11.010	12.970	10.81	15.57	2.530
Agreement with					
population chron.	0.91	0.928	0.915	0.940	0.710
% var. 1st.					
Eigenvector	47.830	51.790	49.29	46.430	38.810
No trees	14	14	13	20	8
No radii	19	18	18	27	8
Signal/trees	0.786	0.928	0.831	0.778	0.420
Signal/radii	0.579	0.720	0.875	0.576	0.310
Correlation coeff.					
chron./log10 prec.	0.717	0.692	0.743	0.699	0.542
Fit:					
Standard error					
of the residuals	0.1542	0.1599	0.1483	0.1583	0.1851
% expl. variance	32.20	28.95	40.77	28.90	11.90

on each reconstruction using the VERIFY program (Gordon and LeDuc, 1981). These tests are: 1) the correlation coefficient, 2) the reduction of error (RE), 3) the t-test, 4) the count of agreement in sign and 5) the chi-square on a 2 x 2 contingency table.

The correlation coefficient measures the strength of the linear relationship between two variables. It is, however, insensitive to any difference in scales of the two data sets and very sensitive to long-term trends. The effects of trends can be dealt with by calculating the correlation coefficient from the first difference of the two data sets. This correlation coefficient measures the similarity of high-frequency co-variance.

Since the regression model has been "tuned" to fit the data over the calibration period, the correlation coefficient is expected to be lower over the verification period; but if the model is good, the correlation must still preserve the statistical significance.

The reduction of error, RE (Kuzbach and Guetter, 1980; Lorenz, 1956) is a rigorous test of the reliability of the estimated parameter. If RE is calculated for the calibration period, its value is the square value of the correlation coefficient. During the verification period, RE can range from − OO to 1. An RE of zero occurs when the regression estimate is just as good as the trivial persistence estimate, so the reconstructed estimates are not better than a reconstruction made using the predictant mean of the calibration period. Values of RE > 0 are a proof of the usefulness of the regression estimate.

No formal significance test exists for the RE, but simulation experiments (Gordon and LeDuc , 1981) indicate that for n > 10 the approximate 95% level is RE = 0.

The simplest parametric test that might be applied to the problem of verification is the t-test on the means of two time series. By assessing only the means of the series, the test would ignore the important temporal properties of the series.

The sign test is a less sensitive measurement of the reliability of the estimates, and it is performed by counting the number of times that the signs of the observed and estimated series agree or disagree. When it is applied to the first differenced data the inferences concern only the direction of the interannual variation.

A contingency table analysis can reveal non-linear relationships and the chi-square statistic can be used to test the significance of the relationship. The results are shown in Table 3.

All the different reconstructions have passed the tests for the calibration period, however the chronology built with the anomalous cores produced the lowest values of correlation coefficients, RE and t-test. In the verification procedure, this chronology shows, for the late period (1930-1960), no correlation, reduction of the error and t-test significants at the 95% probability level. The results confirm the hypothesis that these cores are anomalous in some way. Besides, the published chronology shows no first difference correlation and no error reduction for the verification period (1930-1960).

On the contrary, the chronology made with the highest related precipitation cores shows the best set of significant statistics for both the calibration and verification periods proving the stability of the results.

4 FIDELITY OF THE FREQUENCY DOMAIN

The verification test (Table 3) indicates that the reconstruction made with the HPPC chronology is the most reliable. However, to check the behaviour of the estimates in the frequency domain, the spectrum, coherence, phase and gain of the observed precipitation series against the reconstructed ones with the HPPT chronology was calculated using the ISP (Interactive Statistical Package) spectrum program. This program filters the data using a Parzen's filter and then calculates the spectra with the fast Fourier transform.

Unfortunately, the precipitation series is quite short and this fact restrains the resolution power of the spectrum. On this basis, only 25 lags were used in the estimated for a set of 92 data (1870-1961).

The comparison of the two spectra (Figure 3) shows a close similarity in all the frequencies lower than 0.372.

The coherence spectrum is the frequency domain analog to the cross correlation coefficient squared between two series. It reveals that the tree-ring reconstruction explains very well the variance of

Table 3. Calibration and verification statistics of the different subsets.

Chronology		N° of Observations	Corr. coefficient (R)	R on 1st. differences	Reduction of error (RE)	Sign Test +	Sign Test −	Sign test on 1st. differences +	Sign test on 1st. differences −	t - test	Chi - squared test	N° of Observations	Corr. coefficient (R)	R on 1st. differences	Reduction of error (RE)	Sign test +	Sign test −	Sign test on +	Sign test on −	t - test	Chi - squared test
				Calibration Period										Independent Period							
Published	a	60	0.61	0.56	0.36	48	12	37	22	3.06	2.83	31	0.35	0.29	−0.05	20	11	14	15	1.92	1.41
	b	60	0.51	0.41	0.25	41	19	33	25	3.28	4.00	31	0.58	0.62	0.31	25	6	18	12	1.97	3.02
Subset PCLF	a	60	0.65	0.60	0.41	46	14	41	18	3.97	1.33	31	0.45	0.39	0.16	21	10	16	14	1.94	1.09
	b	60	0.56	0.47	0.31	42	18	37	22	3.47	1.16	31	0.64	0.47	0.31	24	7	20	10	2.74	1.74
Subset HPPT	a	60	0.70	0.67	0.47	47	13	41	17	3.85	2.16	31	0.50	0.44	0.26	23	8	16	14	1.69	4.32
	b	60	0.65	0.58	0.41	43	17	39	20	3.48	5.16	31	0.66	0.64	0.43	24	7	18	11	2.53	5.93
Subset BADC	a	60	0.46	0.52	0.20	40	20	35	24	2.53	5.83	31	0.14	0.07	−0.92	16	15	18	12	1.05	3.03
	b	60	0.28	0.25	0.05	35	25	36	23	1.80	4.83	31	0.43	0.65	0.11	18	13	18	12·	1.90	6.25
Critical values			0.21	0.21	0.04	23	23	1.67		1.67	5.20		0.30	0.30	0.09	10		9		1.69	5.99

Table 3. Calibration and verification statistics of the different subsets. a) years 1870-1929 in the calibration period, years 1930-1960 in the verification period. b) years 1901-1960 in the calibration period, years 1870-1900 in the verification period.

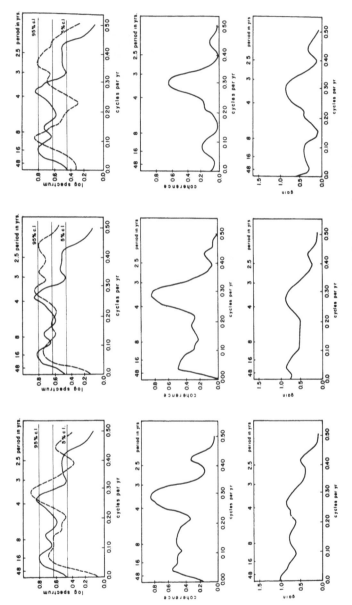

Figure 3. Cospectrum, coherence and gain of Santiago's winter rainfall record and as reconstructed by a) HPPT chronology; b) published chronology (LaMarche et al, 1979); c) BADC chronology. The cospectra solid line indicates the observed precipitation and the dotted line the reconstructed ones. In all cases the number of data is 60 with 25 lags.

the actual winter rainfall data for periods longer
than 2.66 years. These frequency bands have
coherencies up to 45% in the band centered in the
period 3.00 - 4.20 years, it exceeds the 75%.

The gain spectrum reveals how the regression
model has been an effective linear filter in the
transformation of the tree rings in estimates of
the precipitation. A flat gain is desirable because
it indicates that the filter is unbiased in passing
variance in function of the frequencies. The gain
spectrum for the HPPL chronology shows a fairly
unbiased behaviour of the regression model as
linear filter in the wave lengths of 48 to 3.00
years. Variance at very high frequencies is not
passed as well.

Less obvious is the comparison between the
spectra of the published and BADC chronologies and
the precipitation.

Finally, the spectrum of the reconstructed
series was calculated using the same program
(Figure 4). The lag window equals the 16% of the
observations. The oscillatory behaviour seen in the
low pass estimates appears to be quasi-periodic
with a mean period of 8 years, and in high
frequencies, a noticeable peak is shown at the 3.14
years period.

5 DISCUSSION

The main purpose of this paper was to improve the
reconstruction of the winter precipitation of
Santiago de Chile. In order to accomplish this goal
a non-classical methodology was tested. According
to Pittock and Cropper (1982) and Wingley et al.
(1984), more accuracy is obtained in the
reconstruction based on tree-ring chronologies when
more cores are used in the constructions of the
chronology itself because the signal to noise ratio
normally increases.

On the other hand, Peter et al. (1981)
demonstrated the improvement in the climatic signal
contained in a particular chronology using
principal components that allow to examine site
heterogeneities. Using this model, it is possible
to develop chronologies that correlate better with
local climate data than the standard chronology for
a site and which can be tested for time stability
within the framework of the model.

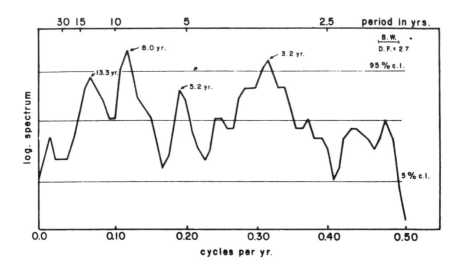

Figure 4. Spectrum of the winter rainfall in Santiago de Chile as reconstructed by chronology HPPT. Number of data: 455. Lags: 60.

In this context, we have sought different approaches to optimize climatic correlation over several criteria of core selections. The results show that the best and simplest procedure was to test the correlation of each core with the climatic parameter and select those which have the higher coefficients. However, all the selection criteria rejected approximately the same set of cores which are regarded as pathological in some way. The chronology made with these cores has failed to produce any reliable reconstruction, even if part of the climatic signal is still present.

The verification test has proved the goodness of the fit of the different models and the HPPT chronology shows the best results of all the comparisons (Figure 5).

In the frequency domain, also the HPPC chronology gives the best gain and coherent spectra.

The analysis of the reconstructed series shows several interesting features. As it could be seen in Figure 6, one long period of droughts is likely to occur between 1270-1450; another drought period appears between 1600-1650, and many others of minor time scale, but anyway longer than any that

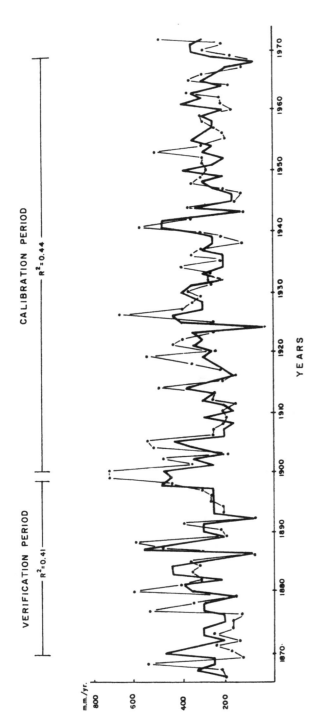

Figure 5. Winter rainfall in Santiago de Chile as reconstructed by chronology HPPT (solid line) and as observed between the years 1870–1960 (dashed line).

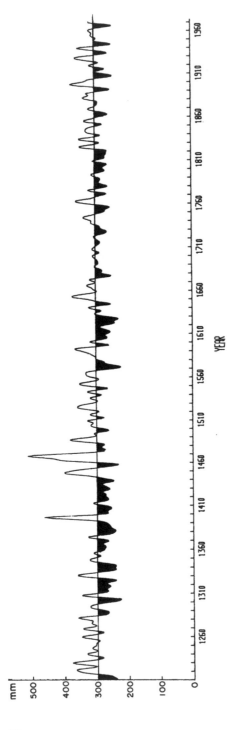

Figure 6. Santiago's winter rainfall series as reconstructed by chronology HPPT, filtered with a 5-years pass digital filter.

occurred since the beginning of the rainfall measurements. Even if a major drought cannot be forecasted, its occurrence can be anticipated as highly probable.

No low frequencies fluctuations seem important in the reconstructed series as it is shown in the spectral estimates variation. This fact could be due in part to the use of the residual series in the development of the chronology, that is, the series were pre-whitened to avoid undesirable effects of differences in the autocorrelation of the individual cores, and probably the procedure causes the loss of part of the long-term fluctuation related to climate. But also, very few long-term variations were found in the precipitation records of central Chile because Santiago usually lies close to the northern edges of storms travelling eastward from the Pacific Ocean and rainfall amounts are highly sensitive to shift in the storm tracks (J. Minetti, personal comm.).

The winter precipitation of central Chile is strongly related to the snowfall in the Central Andes Cordillera (Vargas and Compagnucci, 1985). The economy of a large part of the Central Chile and Argentine oasis is highly dependent on the irrigation water produced by the melting of the mountain snow. Thus, the reconstruction presented here could be regarded as an indirect measure of the availability of irrigation water in those areas.

CONCLUSIONS

The results of this study demonstrate that the statistical techniques are sufficiently developed to produce good quality reconstructions even using one single tree ring chronology. Some simple techniques have proved to be useful to improve the reconstruction by controlling the quality and the characteristics of the signal related to core climate introduced into the chronology.

In other cases, the test of this procedure would be desirable.

The reconstruction of the winter rainfall of Santiago de Chile presented here accounts for approximately the 41% of the total variance over the extended verification period.

ACKNOWLEDGEMENTS

The research for this paper was partially supported by the grant No9780 of CONICET, the National Council of Scientific and Technological Research of Argentina.
The author thanks Richard Holmes and David Mekko of the Tree Ring Laboratory of the University of Arizona for their help and valuable suggestions.

REFERENCES

Benitez, J. 1973. Análisis probabilístico de la variación de la precipitación de Santiago a lo largo del tiempo. Jornadas Científico Tecnológicas "El agua y el futuro regional", pag 59-63, Mendoza.

Cook, E. 1986. On the disaggregation of tree ring series for environmental studies. Proceedings of the International Symposium on Ecological Aspects of Tree Ring Analysis, New York. pp. 522-542.

Cook, E. & Holmes, R.L. 1984. Program ARSTAN Users Manual. Laboratory of Tree Rings Research. University of Arizona, Tucson, Arizona.

C.O.R.F.O. 1971. Pluviometría de Chile, Anexo 1, Estadísticas pluviométricas. Departamento de Recursos Hidraulicos, CORFO, Chile.

Ereño, C.E. & Hoffmann, J.A. 1978. El régimen pluvial de la Cordillera Central. Cuaderno de Geografía No5, Fac. de Filosofia y Letras, UNBA, 36pp.

Fritts, H.; Blasing, T.J.; Hayden, B.P. & Kutzbach, J.E. 1971. Multivariate techniques for specifying tree growth and climate relationships and for reconstructing anomalies in paleoclimate. J. Appl. Met. 10:845-864.

Gordon, A. & Leduc, S.K. 1981. Verification statistics for regression models. Preprints Seventh Conference Probability and Statistics in the Atmospheric Science. Monterrey, Americ. Met. Soc. 129-133.

Holmes, R.L. 1983. Computer assisted quality control in tree ring dating and measurements. Tree Ring Bull. 43:69-75.

Jones, P.D.; Raper, S.C.; Cherry, G.; Goodess, C. & Wigley, T.M. 1986. A grid point surface air temperature data set for the Southern Hemisphere, 1851 - 1984. DOE Tech. Rep. No

27, Carbon Dioxide Research Division, 73p.

Kutzbach, J.E. & Guetter, P.J. 1980. On the design of paleoenvironmental data networks for estimating large scale patterns of climate. **Quat. Res.** 14:169-187.

Lamarche, V. 1975. Potential of tree rings for reconstruction of past climate variations in the Southern Hemisphere. WMO, Proc of the WMO/JAMP, **Symp.** on **Long Term Climatic Fluctuations**, Norwhich, p21-30.

Lamarche, V.; Holmes, R.L.; Dunwiddie, P.W. & Drew, L.G. 1978. Tree Ring Chronologies of the Southern Hemisphere. 2. Chile. Chronology Series V. **Laboratory of Tree Ring Research**, University of Arizona, Tucson, Arizona.

Lamb, H.H. 1972. **Climate: Present, Past and Future.** Vol. 1: Fundamentals and Climate Now. Ed. Methuen & Co London, pp 613.

Lorenz, E.N. 1956. Empirical orthogonal functions and statistical weather prediction. **Statistical Forecasting Project.** Rep. N o 1, MIT, Contract AF19 (604)-1566.

Lough, J. 1983. **Program RESPON Users Manual. Tree Ring Research Laboratory.** University of Arizona, Tucson, Arizona.

Peter, K.; Jacoby, G.C. & Cook, E. 1981. Principal Components Analysis of Tree Ring Sites. **Tree Ring Bull.** 41:1-19.

Pittock, A. & Cropper, J. 1982. Climatic reconstructions from tree rings. In: **Climate from tree rings**, Ed. M.K.Hughes, P.M. Kelly, J.R. Pilcher and V.C. Lamarche, Cambridge Univ. Press. pp. 62-66.

Quinn, W.V. & Neal, V.T. 1982. Long Term Variations on the Southern Oscillation, El Niño and Chilean Subtropical Rainfall. **School of Oceanography**, Oregon State Univ. Oregon.

Taulis, E. 1938. De la distribution des pluies au Chili. Materiaux pour l'étude des calamites. 33:3-20. Geneva (Soc. Geogr.).

Vargas, M. & Compagnucci, R.H. 1985. Relaciones del Régimen de Precipitación entre Santiago de Chile y las Series de la Región Cordillerana. **Geoacta** 13(1):81-93.

Wigley, T.M.; Briffa, K.R. & Jones, P.D. 1984. On the average value of correlated time series, with applications in dendroclimatology and hydrometeorology. J. **Clim. Appl. Met.** 23:201-213.

DENIS WIRRMANN
ORSTOM, Bondy, France

PHILIPPE MOURGUIART
University of Bordeaux I, Talence, France

LUIS FERNANDO DE OLIVEIRA ALMEIDA
Convenio UMSA-ORSTOM, La Paz, Bolivia

5

Holocene sedimentology and ostracods distribution in Lake Titicaca – Paleohydrological interpretations

ABSTRACT

Sedimentological analysis from 22 cores taken in Lake Titicaca (Bolivia) in association with a comparative study between the actual distribution of ostracods and the fossil ones, allow us to present the first interpretation curve of the lake level fluctuations during the last 10,000 years and to propose paleohydrological interpretation based on radiocarbon dating. The Holocene period was characterized by a severe lowering of the lake level in comparison with the present one, inducing gypsum precipitation at the maximum of the lake decrease (54 m at least), during the time interval 7500-7000 B.P. It is only after 2200 yr B.P. that the Lake Titicaca reached its present state.

RESUMEN

El análisis sedimentológico de 22 testigos del lago Titicaca (Bolivia) y el estudio comparativo de la distribución actual de los ostrácodos así como de los fósiles encontrados, nos permite presentar la curva interpretativa de las fluctuaciones del nivel lacustre durante los 10.000 últimos años y proponer una interpretación paleohidrológica, basada en dataciones de radiocarbono. El Holoceno está caracterizado por una fuerte bajada del nivel lacustre en comparación con el actual, dando lugar a la precipitación de yeso en la época de descenso máximo (54 m por lo menos) durante el intervalo de tiempo 7500-7000 años B.P. Solamente después de 2200 años B.P. el lago Titicaca ha tomado su aspecto actual.

INTRODUCTION

The Bolivian Altiplano is an endorheic basin which
extends from lat.16o to 22oS   and   from
long.   65o   to   69oW,   around   4000   m
a.s.l.   and covers 200,000 km2 between the Western
and Eastern Cordillera (6500 m high).
       From   north   to   south,   three   principal
lacustrine   areas   occupy   this high plateau   (Figure
1):
- Lake   Titicaca,   at   3809   m   above   sea   level,
covering 8563 km2;
- Lake Poopó,   at 3686 m a.s.l., which extends over
2530 km2;
- Coipasa   and   Uyuni,   a group of dry salt   lakes,
covering 11,000 km2, at a height of 3653 m.
       During   the Late Pleistocene,   two   lacustrine
transgressions occurred (Servant et Fontes,   1978),
especially   marked   in the south of   the   Altiplano
(Figure 1):
- before   22,000 yr B.P.,   the lacustrine Minchin
phase   produced   a lake of 63,000 km2   over   Poopó-
Coipasa-Uyuni.   The higher water level,   underlined
in   the landscape by stromatolites,   was located at
104 m above the present surface of the Uyuni Salar,
and the water contained 30 to 130 gr/1 of dissolved
salts (Roux et Servant-Vildary,   1987).   Around the
Titicaca,   the   Minchin phase is recognized in   the
northern and western coasts,   at 10-15 m above   the
present lake level;
- after 22,000 and before 13,000 yr B.P.,   a strong
lowering   occurred,   with precipitation of a saline
crust in the southern basin. No dates are available
for Titicaca;
- from   13,000 to 10,500 yr B.P.,   the   lacustrine
Tauca phase produced a paleolake of 43,000 km2 over
Poopó-Coipasa-Uyuni.   The water was salty (20 to 70
g/1),   but several   inflows of   fresh-water   are
registered. Some sedimentological relics, 5 m above
the   present level of Lake Titicaca at the south of
this basin,   are attributed to this stage;
- after   10,500 yr B.P.,   the southern   basin   was
characterized   by very low lacustrine levels   and
brackish   waters,   giving rise to   the   evaporitic
deposits   at   Coipasa-Uyuni and turning   the   Poopó
into a very shallow lake (less than 6 m deep).
       The   study of   level fluctuations of   the
Holocene lakes was initiated in 1982,   using Zullig
corer and Mackereth corer in lake Titicaca. The job
is   developed   through   the   GEOCIT   program

90

(Géodynamique du Climat Tropical) started by the
ORSTOM (Institut Français de Recherche Scientifique
pour le Développement en Coopération) and is
conducted in Bolivia in collaboration with the UMSA
(Universidad Mayor de San Andrés) of La Paz.
    This paper will be devoted to a
detailed analysis of four reference cores and to
the presentation of the paleohydrological
interpretations based on the results obtained in
all the cores studied as yet.

REGIONAL BACKGROUND - PRESENT CONDITIONS

The origin of lake Titicaca is still not well
established and subject to discussion. Lavenu
(1981) speaks of the filling of a tectonic
depression produced by a distension phase during
the Late Tertiary. Later on, during the Pleistocene
various lacustrine episodes occurred. The first
ones correspond to the phases named as lake Mataro
and lake Cabana attributed to the Early Pleistocene
and Middle Pleistocene, respectively. The
corresponding deposits are well recognized on the
northern and western sides of the Titicaca basin at
140 m and 90 m above the present lake level (Lavenu
et al., 1984). Then, the Ballivian transgression
generated a paleolake whose level was 50 m higher
than the present one (Ahlfeld, 1972). This episode
is correlated with the ice retreat of the last-but-
one glaciation of the Bolivian Andes called Sorata
(Servant, 1977).
    Lake Titicaca basin extends from 180 km long
and 69 km wide over Bolivia and Peru; it is
elongated according to a NNW-ESE line and presents
two sub-basins (Figure 2):
- at the north, a bigger lake or lake Chuquito,
with a mean depth of 135 m and a maximum of 284 m,
is characterized by a reduced littoral border (less
than 20 m deep) and steep slopes. Lake Chuquito
represents the 84% of the whole Titicaca surface;
- at the east, in the Chua depression, a smaller
lake or southern part, called lake Huiñaymarca, is
characterized by gentle slopes, a mean depth of 9 m
and a maximum depth of about 42 m.
    These two sub-basins are connected through the
Tiquina strait, with a width of 0,8 km and a mean
depth of 40 m. The only effluent, the Desaguadero,
comes up from the small lake and flows
into the lake Poopó to the south, 300 km farther.

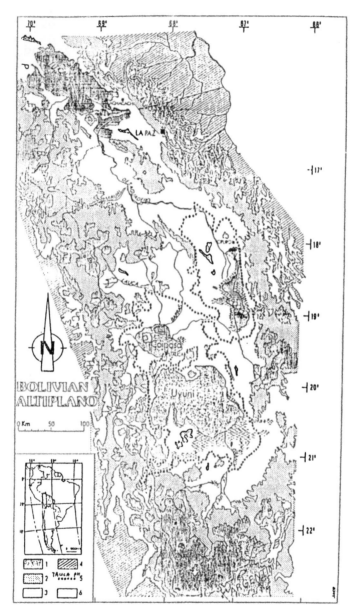

Figure 1. The bolivian Altiplano and the extension of the lacustrine Tauca phase (modified after: Servant et Fontes, 1978).

1	> 4500 m	4	< 3500 m
2	4000-4500	5	limit of paleolake Tauca
3	3500-4000	6	dry salt lakes

92

CLIMATOLOGY AND HYDROLOGY

Lake Titicaca is located in a tableland area of cold and semi-arid climate (Sheriff, 1979). In summer, the maximum temperature is of 21 oC (January-February) and in winter, the lower temperature is less than 5 oC (June-August). Diurnal amplitudes vary from 10oC over the lake to more than 20oC in the surroundings and the rainy season is from October to April (Libermann Cruz, 1987).

Despite a great lacustrine volume of 893 km3, the lake basin has a small extension of 57,340 km2, with gentle slopes inferior to the 10% near the border of the lake (Richerson et al., 1975). Five principal streams bring water to the lake, each one flowing into the big lake. They are the rios Ramis, Huancané, Coati, Ilave and Suchez (Figure 2). The yearly water-inflow attributed to all the tributaries of the lake runs about 8.25 x 10 m3 (Carmouze & Aquize, 1981). Medium annual precipitations range from 980 to 800 mm for the big lake and lake Huiñaymarca, respectively. Direct rainflow over the Titicaca rises to 7.9 x 10 m3 per year, of which a 72% falls during summer and only a 6% during winter (Lazzaro, 1981). Assuming that the lake has no sub-lacustrine springs, its annual water income is 16.75 x 10 m3. These inputs are eliminated by evaporation (90%), infiltration (9%) and the effluent Desaguadero (1%). The water turnover is weak for the big lake (1.6% per year with a residence time for the water of 63.5 years) but higher in the small lake (24% per year and a water residence time of 4.5 years).

Annually, the global water turnover range is 2% (Carmouze & Aquize, 1981), so lake Titicaca is nearly a closed lake. However a considerable difference exists between the big and the small lake, the latter evacuates a 36.5% of its yearly income in comparison with only a 11.6% for the former.

The irregularity of the inflows and the annual distribution of the evaporation induce seasonal fluctuations of lake level to run from 0.6 to 0.8 m around the mean level. The higher levels are registered during April and the lower ones during December. Interannually variations rise to more or less 3 m around the present level, presenting higher amplitudes every 10.6 and 2.4 years (Kunzell and Kessler, 1986).

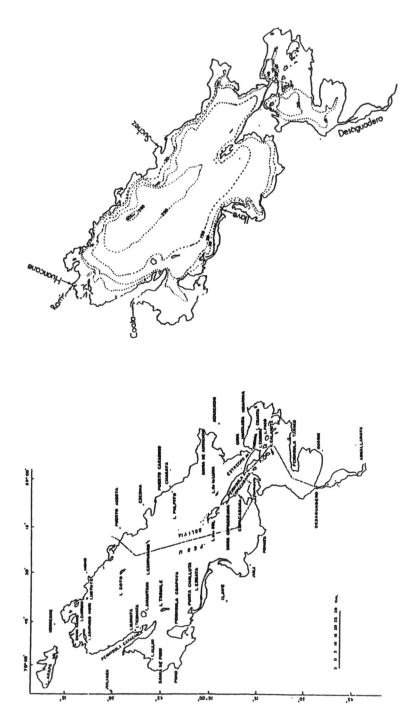

Figure 2. Lake Titicaca: toponomy and bathymetry, main tributaries.

94

The main fluvial input is due to rivers coming
from volcanic and sedimentary formations with high
salinities (5 to 20 mM/l) and a bicarbonated-
sulfate-chlorine facies; the rest is due to
affluents coming from crystralline formations with
low salinities (5 mM/l) and a bicarbonate-calco-
sodic facies. The river waters are concentrated on
the lake by a factor of 3.4; the mean annual salt
output results from infiltration (40%), chemical
sedimentation (28.5%) and the effluent Desaguadero
(11.5%); the rest is stocked in the lake. The
yearly salt turnover varies from 0.2 to 0.5%
according to the major elements; consequently, time
fluctuation in the hydrochemistry of the lake is
very weak. The salt concentration in the lake water
is about 1.2 gr/l and the water is characterized by
a chlorine-sodic facies (Carmouze et al., 1981).
From the surface to the bottom of the lake,
the mean water temperature varies from 18oC to
8 oC and the pH ranges from 8.8 to 7.2,
with a rare extreme value rising to 9.38 in the
Guaqui Bay - small lake - (Quintanilla et al.,
1987; Iltis, 1987). From October to June, a
stratification is observed in the big lake with
a thermocline at 40-70 m in March and deeper - 100
m - in June (Carmouze et al., 1984; Quintanilla et
al., 1987) after the big lake turn isotherm. The
small lake is classified as a polymictic warm lake
differing from the Chua depression which is a
monomictic warm area from August to September and
after isotherm, as the big lake (Richerson et al.,
1977; Lazzaro, 1981).

LIMNOLOGY

Bathymetry and local bottom relief control the
distribution of the aquatic life and hence the
sedimentological characteristics of the lake. The
submerged vegetation, more developed in the small
lake than in the big one, plays a predominant role
because it is the preferred place for benthic or
planktonic species and aquatic vertebrates
development, encountered as fossils in the
sediments (shells, diatoms, sponge spicules,
macrophytes remains and so on), and so it is
indirectly responsible for the sedimentological
facies distribution.
Collot et al. (1983) differentiate five main
levels of aquatic vegetation according to the plant
determination (Guerlesquin, 1981):

- in the shoreline area - up to 0.2 m deep - the macrophytes association is represented by Lilaecopsis and Hydrocotyle;
- from 0.2 to 2.5 m, **Myriophyllum** and **Elodea** in association with other less representative and abundant species (Potamogeton, Zannichellia, Ruppia and Sciaromiun) are characteristic of this depth interval;
- between 2.5 m and 4.5 m, Schoenoplectus tatora, the famous totora, is very well developed. It represents the 30% of the entire subaquatic plant surface in the small lake;
- in lake Huiñaymarca, the chara (**Chara spp.**, **Lamprothamnium**) composes the 60% of the global macrophytes population and grows between 4.5 and 7.5 m excluding mainly the colonization by other species. It can be rarely recognized up to 9.5 m;
- after 7.5 m and up to 12 m, **Potamogeton** prevails.

Molluscs, crustaceous, diatoms and sponges are the most important groups living in this habitat. Molluscs are represented by three main epiphyte species of gastropods and one limnicolus species of bivalve (Haas, 1955). The shells are made of pure aragonite which does not suffer any mineralogical transformation during burial (Wirrmann, 1982).

The modern ostracod distribution is directly controlled by the combination of two factors (Mourguiart et al., 1986; Mourguiart, 1987): the ionic composition of the lake water and of its tributaries, and the equilibrium between carbonates and organic matter. According to these facts, the distribution of the ostracod fauna is related to the bathymetrical zonation of the aquatic vegetation. Six main associations are differentiated:
- from the shoreline up to 2.5 m, the biotope A is represented by very few planktonic species adapted to turbid waters and to the seasonal lake level fluctuations. The principal species are: **Chamydotheca**, **Ilyocypris** and **Herpetocypris**. The fossil tests are very rare, due to the fragility of the shells and to the high level of energy found in this area;
- between 2.5 and 4.5 m, the totora provides peculiar characteristics such as the lack of dissolved oxygen at the bottom (Polunin, 1984) and high bacterial activity inducing an increase in CO_2 and acidifying the pH at the water-sediment interphase and the unfavourable conditions to zoobenthos development;

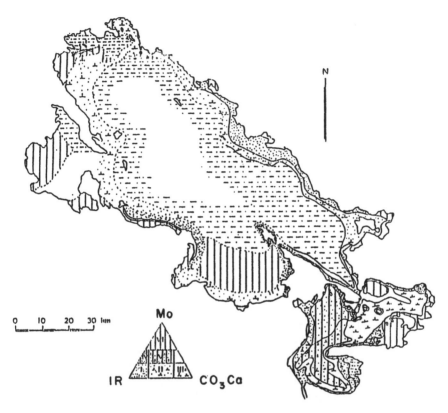

Figure 3. Surficial sediments repartition in the
lake Titicaca (after: Boulanger et al., 1981).

I   Detritic sediments
II   Detritic and carbonated sediments
III Carbonated sediments
IV   Organic and detritic sediments
V   Organic, detritic and carbonated sediments
VI   Organic sediments

- in the interval 4.5 m - 7.5 m, a sparse
population of benthic ostracods (100 valves per
cm3) is principally represented by Limnocythere
charaensis n. sp. (morphe B 7), Limnocythere
titicaca and Candonopsis sp. A.. These three
species are dominant in turn without following a
bathymetrical rule but always represent the 90% of
the global ostracod fauna; the other 10% is
composed by Darwinula sp., Cypridopsis aff.
huaronensis and Limnocythere sp. Y.. The common

characteristic of these species is a very thin test
and undersized height;
- the biotope D is linked to the **Potamogeton** zone
and corresponds to a very high proliferation of
ostracods (the density ranges from 5 to 25 g/m2)
represented by: **Limnocythere gr.** A polymorphic,
Limnocythere gr. B polymorphic and **Limnocythere
titicaca** associated with **Candonopsis sp.** A and **sp.**
B, **Darwinula** and **Amphicypris**;
- below 12.5 m and down to 20 m, the lack of
organic matter leads to a lowering of the faunal
density and variety; the ostracods are represented
by **Candonopsis sp.** A. and **Limnocythere gr.** A and
gr. B.;
- further down, in the deepest parts of the lake,
the ostracod fauna is characterized by a very low
density. The 90% of the species is represented by
Candonopsis sp. A associated with **Limnocythere
sp.** A1 and **Darwinula**. This fauna is only found in
living forms, due to the acid pH at this depth,
which prevents the tests conservation in the
sediments.

Locally, this distribution is disturbed by the
influence of debris inflow and of water enriched in
Na+ or Cl-   conducing to a contraction
or to an enlargement of the biotopes limits, more
sensible for the three latter zones.

The determination of the present brackish
water fauna was made in the area of lake Poopó and
its neighbouring temporal pounds. The adaptation to
hard life conditions (high variability of water
level, high dissolved salt concentrations,
dehydratation and so on) leads to the development
of a special fauna presenting short life cycles and
represented by **Cypridopsis sp.**, **Potamocypris sp.**,
Amphicypris sp. and especially by **Limnocythere
bradburyi** which can be considered as the best
paleoecological marker of such environments.

The distribution of surficial sediments
is strongly correlated to the bathymetry (Boulange
et al., 1981) because most deposits at the bottom
have a biogenetic autochthonous origin (Figure 3).
The detritical sediments - except the finest ones -
are located at the mouth of the rivers and on the
shores, due to the filtration role assumed by the
macrophytes. Organic-calcareous deposits are
representative of shallower zones (up to 15 m) in
relation to a high macrophytes colonization. In
the deepest areas, the sediments are composed of
very fine detritical grains and of clays mixed with

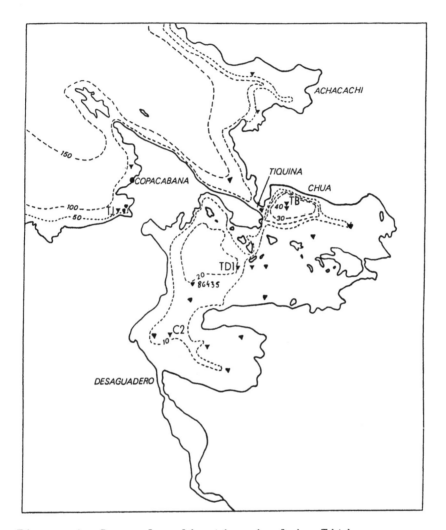

Figure 4. Cores Localization in lake Titicaca.

organic matter resulting from the plankton and
macrophytes decomposition. In the big lake, the
calcareous sediments present a wider distribution
due to the abrupt slopes.

CONCLUSIONS

The limnological and physical properties in lake
Titicaca differentiate three main areas, each one

characterized by a specific environment, reflecting the hydrological and bathymetrical influences, which are well registered by the nature of the sediments and the ostracod fauna:
- the big lake and the Chua depression in lake Huiñaymarca are the more homogeneous parts of lake Titicaca;
- the rest of the small lake and the principal bays of lake Chuquito present a similar and moderate response to the environmental fluctuations and
- the shore shows specific limnological characteristics, recording the stronger variations due to seasonal and interannual changes of the environment.

SAMPLING

Based on previous results concerning the surficial deposits and the first samples taken with a Zullig corer en 1982, three transects crossing the main parts of the small lake and the bay of Achacachi and Copacabana - Yunguyo in the big lake, were chosen for the Mackereth sampling (Figure 4).
        In this paper, only four cores will be analyzed in detail (for more information about the other cores see Oliveira-Almeida, 1986; Mourguiart, 1987; Wirrmann, 1982 and 1987). The lithological description is based on a morphological study (texture, faunal contents, colour codification with the Munsell Coil Colour Charts, 1975) and the different phases for each core are presented, according to the following graphic representation:

Plastic clay	Shells
Compact clay	Shell Fragments
Very fine clay	Plant remains
Carbonated crust	Calcified macrophytes
Paleosol	Organic laminations Carbonated concretions
Sand	Gypsum

Water content, coarse fraction content (larger than 63 microns, separated by sieving under distilled water), carbonate and organic carbon content are expressed in percent (%) of the dry weight of sediment.

RESULTS

1)   The core TD1

    Sampling   zone: western central part of   lake
                   Huiñaymarca   (Figure 4)
    Length of the core: 537,5 cm
    Depth: 19 m

    A   morphologic   analysis   (Wirrmann   and   de
Oliveira Almeida, 1987) differentiates six main
lithologic units (Figure 5):
- from 537.5 to 332.5 cm: a very fine clay sediment
- mean colour 10YR 3,5/1 - including millimetrical
black concretions;
- from 332.5 to 181 cm: a plastic clay, colour N4/;
- from 181 to 155 cm: a compact, homogeneous fine
mud - mean colour 5GY4/1 to N4/;
- from 155 to 141,5 cm: a silty-argillaceous
sediment (colour N5/ to N4/) including gypsum
crystals;
- from 141.5 to 87 cm: a gelatinous mud with
diffuse banding (mean colour 5Y 5/1) poor in
shells, very rich in gyrogonites and calcified
macrophytes remains;
- from 87 to 0 cm: a gelatinous-organic mud with
diffuse banding (medium colour 5Y 3/1) containing
calcified macrophytes remains and very rich in
gastropods and ostracod shells.
These six lithological   units have been regrouped
in four major sedimentological sequences, according
to   the vertical distribution of the physical
parameters of the bulk sediment. Further, the
observation   of   the   coarse   fraction   and   the
distribution of the ostracods allow to precise the
lake level for each sequence (Figure 5, 6 and Table
1):
- sequence IV: 537.5 - 155 cm, corresponds to very
fine and azoic deposits (less than 1% of silty
fraction), containing 100% of water and 1 to 5% of
organic carbon, with 10% of carbonates. The bulk of
the coarse fraction is composed of syngenetic
concretions, more abundant at the top of this

101

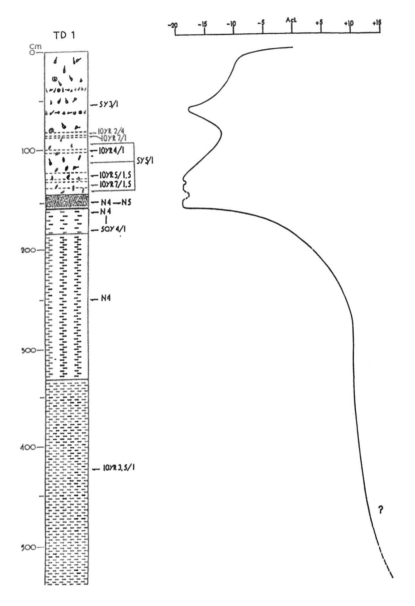

Figure 5. Core TD 1: lithology and paleobathymetry interpretation.

sequence, in association with fine detritical grains (quartz, feldspar, etc.). From the base to the top, the concretions are vivianite, greigite and pyrite. From 250 cm, diatoms and sponge spicules are very abundant and the last ones compose the 99% of the coarse fraction between 170-160 cm.

The base of this sequence is representative of a sedimentary epoch with a high water level, which gave rise to a strong biogenic production along the shores of the lake, inducing vivianite formation in the central hollow of the basin. From 170 cm, the water level is going down rapidly (presence of diatoms and sponge spicules). One 14C dating obtained from carbonates analysis of the bulk sediment for the level 158-157 cm yielded an age of 7700 +/- 230 yr BP; so the base of this sequence might represent the final of Tauca phase, hypothesis confirmed by the age of 9620 +/- 90 yr BP obtained on the neighbouring core TD (Oliveira Almeida, 1986) for an organic level, at 35 cm below the evaporitic deposits;
- sequence III: 155-132 cm, is characterized by evaporitic grains in the shape of lenticular gypsum (the basal one is corroded) and of small "rose des sables", more abundant between 155 and 148 cm. The amount of coarse fraction is high (up to 80%). The ostracod fauna is represented by **Limnocythere bradburyi**, a characteristic species of brackish water.

In comparison with the precedent stage, this sequence reflects a strong lowering of lake level, at least 18 m, perhaps with small oscillations while the basal gypsum crystals are dissolved;
- sequence II: 132-84,5 cm, characterized by carbonated sediments (50 to 90% of $CaCO_3$) including a lot of calcified macrophytes remains and turbid salt water ostracods (**Limnocythere bradburyi** and **Cypridopsis**) is representative of the rising of the lake level. Below 110 cm, the ostracod fauna characterizes oligosaline to fresh waters and the gastropods are more abundant. The depth never exceeded 7-9 m (none of the chara grows deeper, Collot et al., 1983) and more probably was established around 5 m at the top of this sequence;
- sequence I: 84,5 - 0 cm, corresponds to softer sediments, richer in water (medium content 50%), containing 30-40% of $CaCO_3$ and 10% of organic carbon. Most of the silty fraction is composed by shells. The ostracods are representative of

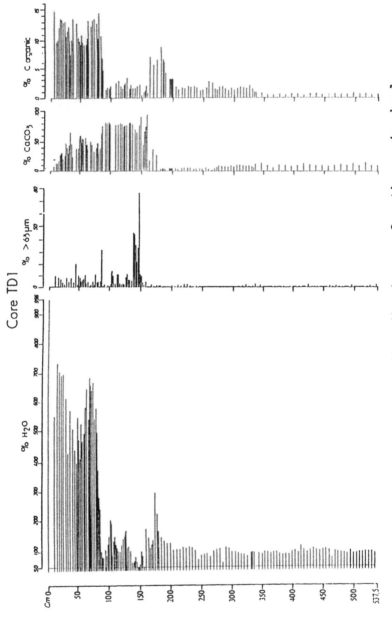

Figure 6. Core TD 1: distribution of the physical parameters (after: Wirrmann and Oliveira Almeida, 1987)

Table 1. Distribution of the main components of the coarse fraction in Core TD-1 (After Wirrmann and Oliveira Almeida, 1987).

Distribution of the main components of the coarse fraction in Core TD-1 (After WIRRMANN and OLIVEIRA ALMEIDA, 1987)

Sequences	Level (cm)	Ostracods	Gastropods	Calcified macrophyte remains	Gyrogonites	Diatoms	Sponges spicules	Others
I	0-84.5	-	+++	+	-	t	-	-
II	84.5-132	-	+	++	+++	t	±	-
III	132-155	-	-	-	-	+	-	gypsum +++
IV	155	-	-	-	-	++ from 250 cm	++ from 170 cm	+++ pyrite greigite vivianite detrital grains

t = trace; + = present; ++ = abundant; +++ = very abundant.

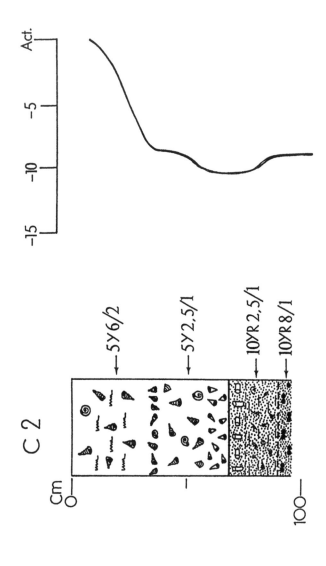

Figure 7. Core C2: lithology and paleobathymetry interpretation.

permanent oligohaline waters (**Limnocythere**, **Darwinula**, **Candonopsis**). Between 65-60 cm, the absence of ostracods attributed to a lowering of the lake level and perhaps to a short emersion time. After 60 cm and up to 20 cm, the ostracod association is dominated by **Limnocythere sp. A** suggesting the establishment of two few contrasted seasons in the lake; medium depth was established around 8 m. Finally, the rising of the lake level up to the present one, is registered from the last 20 cm.

2) The core C2

Sampling zone: south of the western central
hollow in the small lake
(Figure 4)
Length of the core: 95 cm
Depth: 13,5 m

Three main lithological units (Wirrmann and de Oliveira Almeida, 1987) are recognized in this Zullig core (Figure 7):
- from 95 to 90 cm: a white indurated carbonated sediment (10YR 8/1), with a coarse texture containing calcareous concretions (smaller than 0,5 cm) and with few gastropods and ostracod shells at the top;
- from 90 to 68 cm : a black deposit (10YR 2,5/1) partly indurated, very rich in calcified macrophytes remains at the top and containing many shells fragments between 90 and 75 cm; diatoms are well represented in this unit; detrital grains (quartz, feldspar) and clays are abundant. The contact with the lower adjacent unit is very sharp;
- from 68 to 0 cm: a gelatinous mud with diffuse banding (5Y 6/2 to 5Y 2,5/1), very rich in gastropods and ostracod shells. Organic laminations are abundant between 25 and 7 cm.
According to the paleontological data based on the diatoms analysis (Servant-Vildary in Wirrmann and de Oliveira Almeida, 1987) and on the ostracods distribution three sequences are defined:
- sequence III: 85-75 cm, corresponds to the white basal sediments and to the lower part of the black deposits. The ostracods are probably removed, since the tests present signals of disturbation. At the top of this sequence a terrestrial gastropod was found. For the 10 basal centimeters, the diatoms

are representative of shallow fresh water (less
than 3 m deep and less than 3 gr/l of dissolved
salts), but from 80 cm to the top, a drastic change
takes place: the diatoms correspond to shallow
hypersaline brines;
- sequence II: 75-50 cm, is represented by the
upper black deposits and the basal, organic-
calcareous sediments. At the outset, the waters
rise to higher concentrations of dissolved salts
(more than 40 gr/l). The ostracod **Limnocythere
bradburyi** and the diatoms indicate that the depth
was established around 2 m. After 58 cm, the
diatoms and ostracod associations reflect a gentle
rising of the lake level and the waters contained
around 10-20 gr/l of dissolved salts. According to
the 14C dates obtained on the bulk sediment of the
intervals 80-70 and 70-60 cm, which are
respectively 5325 +/- 395 yr BP and 3650 +/- 330 yr
BP; this low lake level remained during two
thousand years approximately;
- sequence I: 50-0 cm, like the top of the organic-
calcareous mud is representative of the gradual
rising of the water. The salinity is established
around 5 gr/l but a few levels (especially between
35-32 cm) suggest inflow of salty waters. The depth
never exceeded 5 m up to the level of 30 cm, and
after the notable increase of **Candonopsis sp. A**
reflects the influence of the present conditions.

3) The core TB

    Sampling area: Chua depression - small lake
                        (Figure 4)
    Length of the core: 495 cm
    Depth: 39 m

    Through the morphological description four
main lithologic units are individualized (Figure
8):
- from 495 to 441 cm: an azoic plastic clay, colour
N4/;
- from 441 to 406 cm: a clay sediment (colour N4/)
presenting a foliated structure in the 10 basal
centimeters. Then, a polygonal structured deposit,
containing a lot of plant remains, is attributed to
a paleosol formation. At 430 cm, a few shells are
present;
- from 406 to 271 cm: a compact, homogeneous clay
(colour 5Y 4/1) at the bottom giving range to silty

deposits and to silty-argillaceous sediments rich in plant remains (colour N4/). The upper 10 cm contain rare shells;
- from 271 to 0 cm: a gelatinous, organic-calcareous mud, rich in shells (fragmentary and complete) with diffuse banding (colours 5Y 3/1,5 to 7,5Y 3/2).

According to the vertical distribution of the physical parameters and to the paleontological results, these four lithological units have been regrouped in three sequences (Figure 9, Table II):
- sequence III: 495-441 cm, corresponds to fine deposits (less than 0.1% of coarse fraction) with a medium water content of 70%. The silty fraction is composed by very fine detrital grains (quartz, feldspar, micas) associated with greigite and vivianite. This sequence suggests a depositional environment deeper than the present one;
- sequence II: 441-271 cm, is represented by silty deposits (10 to 93% of coarse fraction) including two levels (416-385 and 286-280 cm) containing less than 1% of coarse fraction. Besides quartz, feldspar and micas, the coarse level (370-311 cm) is constituted by volcanic elements (especially brown tourmaline, garnet, zircon and andalousite). From 311 cm a lot of plant remains are present.

These deposits are attributed to a sedimentation in a shallower area than the present one, and perhaps with a short emersion phase (interval 441-430 cm similar to a paleosol). Then, the shells appeared at the same time than calcified macrophytes remains (level 430-420 cm). The depth did not overpass 10 m, but the poor state of conservation of the ostracods, hinders to precise the paleohydrological conditions;
- sequence I: 271-0 cm, is characterized by a gelatinous-calcareous mud containing less than 5% of coarse fraction and around 250-300% of water. The bulk silty fraction is composed by shells (gastropods and ostracods) and few detrital grains (especially micas). Associated with ostracods of oligosaline and shallow waters, diatoms representative of salt-water are present in the interval 271-243 cm. From 160 cm, the presence of Limnocythere titicaca and L. sp. B1 reflects a deeper lake level. After 106 cm, the rising of lake level is more and more important and the establishment of the present level is reached from the upper 30 cm. Small lake level fluctuations are registered, especially between 120-75 cm, according to the ostracod fauna.

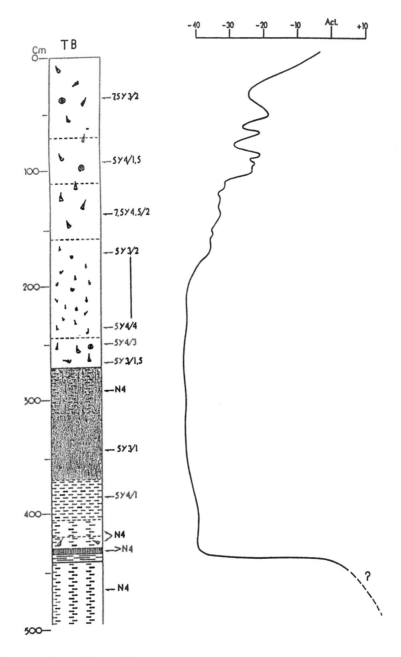

Figure 8. Core TB: lithology and paleobathymetry interpretation.

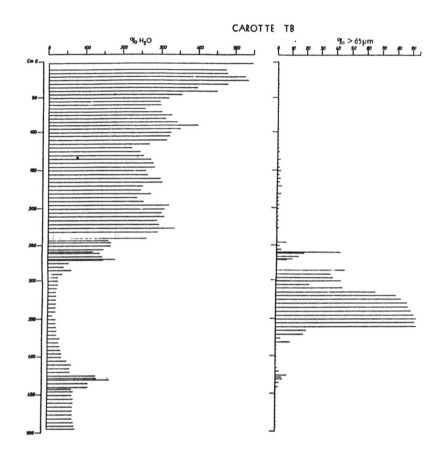

Figure 9. Core TB: distribution of the physical parameters.

4) The core TJ

Sampling area: Yunguyo bay - big lake (Figure 4)
Length of the core: 405 cm
Depth: 50 m

This core, the longest one obtained in the big lake, presents 4 main lithological units (Figure 10):
- from 405 to 394, 5 cm: evaporitic deposits (lenticular gypsum) including plant remains. The colour varies from 5Y 5/1 to 5Y 3,5/2;
- from 394.5 to 320 cm: a sandy-argillaceous

Table 2. Distribution of the main components of the coarse fraction in Core TB

Distribution of the main components of the coarse fraction in Core TB

Sequences	Level (cm)	Ostracods	Gastropods	Calcified macrophyte remains	Gyrogonites	Diatoms	Sponges spicules	Others
I	0-156	+++ from 20 cm	++ to +	-	-	-	t 115-136	+ carbonated fragments
	156-243	+ to t	+++	-	-	-	t	t detritics
	243-256	+++ to +	+ to ++	-	-	t	t	t carb. fragments
	256-271	t	+	+ to +++	t	t	t	t detritics
II	271-281	+ to +++	+++ to +	-	-	-	-	t pyrite detritics
	281-420	t to -	t to -	-	-	-	t	+++ detritics
	420-441	t	+ to -	t	t	-	-	t plant remains
								+++ detritics and greigite
III	441-476	-	-	-	-	t	-	+++ detritics and greigite
	476-495	-	-	-	-	-	-	+++ detritics
								+ carb. frag.
								+ greigite

sediment - colour 5Y 6/1 to 5Y 7/1 - containing shells;
- from 320 to 100 cm: a compact sandy deposit, very rich in shells, with diffuse banding (5Y 5/2 to 2,5Y 7/2);
- from 100 to 0 cm: a gelatinous organic-calcareous mud, very soft, including a lot of shells (colour 5Y 2,5/1 to 5Y 6/3).

Despite that the distribution of the physical parameters (Figure 11) and of the components of the coarse fraction do not present notorious variation along the core (except for the uppermost meter), these four lithological units have been regrouped in three main sedimentological sequences:
- sequence I: 405-394,5 cm, represented by evaporitic deposits and characterized by the irregularity of water volumes (60 to 170%), coarse fraction (6 to 57%) and carbonates (10 to 80%) reflects a very shallow deposition zone, established around 2 m;
- sequence II: 394.5-100 cm, includes two kinds of deposits characterized by high fluctuations of density and/or variety of ostracod fauna. From the bottom of this sequence to 320 cm, there is a calcareous sediment (80% of $CaCO_3$ and less than 1% of organic carbon) which contains around 70% of water and 25% of coarse fraction. The first 20 cm correspond to the setting in water of the basin with oligosaline waters. Then, the lake level rises up to 5-8 m but with a special response of the ostracods from genus *Limnocythere* (Mourguiart, 1987). They present a very high variability of size and of the test architecture (reticulation, macronodation, microcones formation). In comparison with the present conditions their polymorphism is multiplicated by 3; this argues for a very strong seasonal contrast between summer and winter.

From 320 cm to the top of this sequence, the deposits are coarser (50% of fraction larger than 63 microns) and richer in shells. Contents of water vary around 80%; the carbonates amounts do not exceed the 50% and there is less than 1% of organic carbon. The water depth shows gentle fluctuations around 5 m with alternative periods of high seasonal contrast and no contrast at all.

This sequence took place during 2600 yr at least, according to 14C dates of the bulk sediments for the levels 393-390 m and 110-107 cm, which yielded the ages of 7250 (+190, -180) yr B.P. and 4600 (+430, -410) yr B.P., respectively;

113

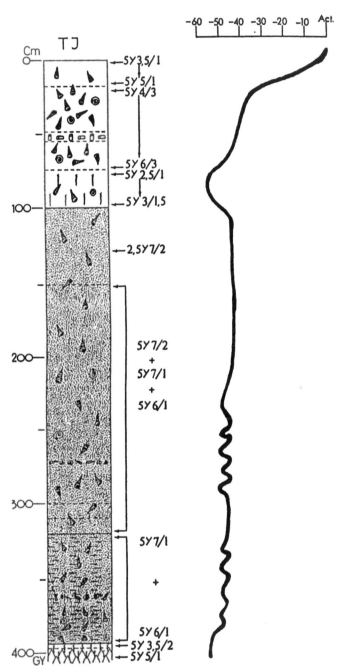

Figure 10. Core TJ: lithology and paleobathymetry interpretation.

- sequence I: 100-0 cm, corresponds to the upper organic-calcareous mud, containing 50% of carbonates, 2% of organic carbon, an average of 150% of water and a proportion of coarse fraction ranging from 1 to 80%. In the lower 25 cm, ostracod species of oligosaline waters are substituted by species of mesosaline and shallow waters like **Cypridopsis sp.** and Limnocythere bradburyi. This lowering of the water level probably implies a short emersion time. Later on, deepening and desalting of the water occurred. From 30 cm, ostracods representative of deeper environment appear (**Limnocythere gr. A. and L. gr. N.**); the present lake level is established.

5) Complementary data

Core 86435

Sampled 18 m below water in the central western part of the small lake (Figure 4), this core shows two lithological units:
- from 46 to 43 cm: a white indurated calcareous sediment, containing a few shells;
- from 43 to 0 cm: a gelatinous, organic-calcareous mud, very rich in shells.
A 14C dating obtained from analysis of the white bulk basal sediment yielded an age of 2200 (+430, -400) yr B.P. and hence gives the age of the beginning of the ultimate rising of the lake level, underlined by the presence of **Cyprideis sp.**, reflecting inflows of waters enriched in dissolved salt at the outset of the filling of the basin.

Core S5

Taken in the western hollow of lake Huiñaymarca (Figure 4), below 17 m of depth, this core presents three lithological units (Wirrmann, 1982):
- from 100 to 90 cm: a compact indurated sediment – colour 5Y 5/2,5 – containing small calcareous concretions (0,5 cm);
- from 90 to 40 cm: a coarse calcareous sediment, compacted, very rich in shells, colour 5Y 4,5/2;
- from 40 to 0 cm: a gelatinous, organic-calcareous mud (5Y 3/2) including a lot of shell fragments.
One 14C dating obtained on the bulk sediment for the interval 97,5 - 93,5 cm yielded an age of

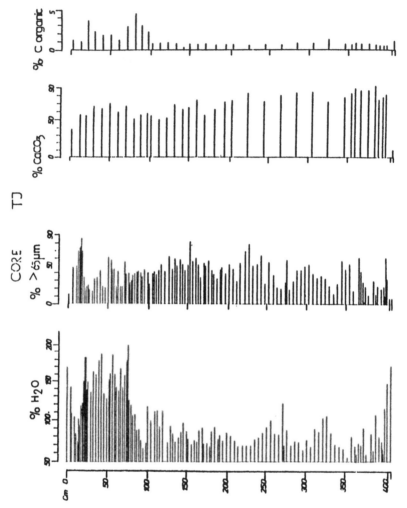

Figure 11. Core TJ: distribution of the physical parameters.

116

5080 +/- 110 yr B.P. Ostracods and diatoms present in this level are representative of polysaline shallow waters, which confirms the interpretation proposed for the cores C2, and TD1.

Core B1

In the south of the Chua depression, below 5.6 m (Figure 4), this core presents an homogeneous lithology along its 140 cm (Wirrmann, 1982) characterized by an organic-calcareous mud (colour 5Y 6/3), very rich in shells and calcified macrophytes remains. An age of 280 +/- 120 yr B.P., obtained on the bulk sediment for the interval 137.5 - 132 cm, confirms that the present conditions have been established very recently and gives an indication about the rate of sedimentation in the Characeae zone: 5 mm/year.

CONCLUSIONS

A comparative analysis of all the cores, based on the different sedimentary sequences recognized, their respective paleobathymetry, the 14C datings, and the absence of turbid perturbations, allow us to present the following correlations (Figure 12):
- the basal sequences of cores TB and TD1, characterized by azoic plastic clays, reflect a period of high lacustrine level, older than 7700 yr B.P.; in the case of core TD1, the basal sediments might form part of the late Tauca lacustrine deposits;
- furthermore, the sedimentation is represented by evaporitic facies (gypsum in cores TD1 and TJ), by deposits of very shallow environment (core C2) or by paleosol structure (core TB) and corresponds to the maximum lowering of the lake level which took place till 7250 yr B.P.;
- later, in all the cores, the sedimentation reflects a gradual rising of the water level with minor fluctuations;
- finally, the filling of the lake concludes with the establishment of the present level, which is registered from the upper 35-20 cm depending upon the cores.
    According to the main areas in the Titicaca - the big lake, the Chua depression and the western central hollow of the small lake -, the rising of

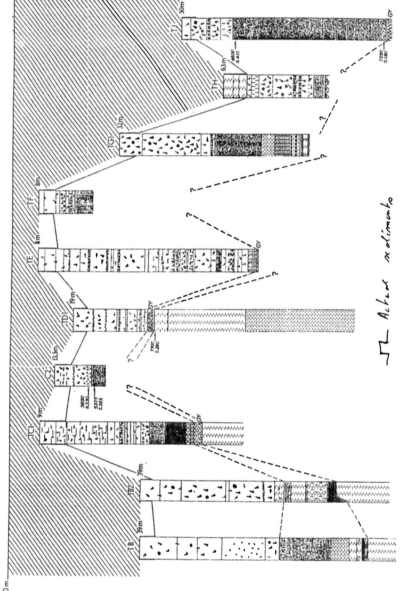

Figure 12. Correlation among the cores.

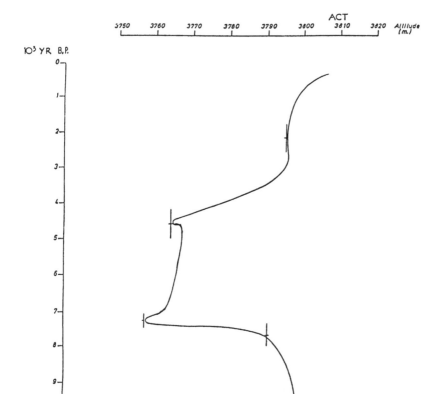

Figure 13. The fluctuation of the lake level during the Holocene.

the water level occurred under different environmental conditions. In the western part of lake Huiñaymarca the beginning of the transgression is marked by hypersaline waters while in the Chua depression the water is oligosaline, as in the big lake.

The synthesis of these results and their paleohydrological interpretation defines five major stages during the Holocene (Figure 13) for the lake Titicaca (Wirrmann & Mourguiart, 1987):
- from 10,500 to 7700 yr B.P., a lacustrine decrease takes place after the Tauca episode

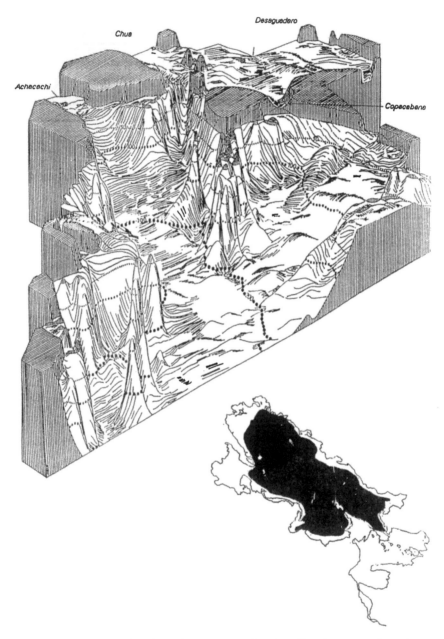

Figure 14. Block diagram of the Titicaca basin, showing the emerged land (▦), the water level at the Holocene maximum lowering (•••50•••) and the corresponding water surface (▰).

(12.500 - 13.000 to 10.500 yr B.P. roughly). From a
deeper stage - around 10 m at least above the
present level -, lake Titicaca registered a severe
lowering, firstly soft and progressive, then more
and more drastic at the end of this period;
- from 7700 to 7000 yr B.P., a lowering of the lake
level, in comparison with the present one, coming
to 54 m at least, with gypsum precipitation on the
bottom of Titicaca. The communication between the
big lake and the small one was cut off. The big
lake lost 227 x 10^9 m3 (26% of its water
volume) and the small lake was dry (Figure 14). The
Titicaca´s surface was reduced to 42%.
- from 7000 to 4700 yr B.P. a gradual rising of the
water level - around 20-40 m below the present
level - is registered according to the cores
position. The water-salt concentration rises to
more than 40 gr/l at the outset and the seasonal
contrast between summer and winter was hardly
marked. The communication between the Chua
depression and the western hollow in the small lake
was still not re-established;
- from 4700 to 2200 yr B.P. approximately, the lake
level has been rising progressively to 10 m below
the present one. Water becomes more and more sweet,
but important inflows of water enriched in Na^+,
Cl- are registered. A short lowering of
the lake level, marked by the development of
mesosaline ostracods in the big lake, occurred
around 4500 yr B.P. The communication between the
small lake and the big one was effective after this
event and the water was fresh from 3500 yr B.P.;
- after 2200 yr B.P., lake Titicaca took its
present state. According to the historical fact
(Ramón Gavilán in Historia de Copacabana, 1621), a
light rising of the water level occurred at the end
of the 16th century and during the beginning of the
17th century.
      The global nature of these changes in lake
Titicaca presents a good correlation with the
regional climatic pattern and with the hydrological
variations registered all over Bolivia (Figure 15).
After the lacustrine stage Tauca - posterior to the
glacial maximum of the last glaciation in the
tropical Andes (Servant & Fontes, 1978) -
attributed to an increase of 30 to 50% of the
rainfall on the Altiplano (Hastenrath & Kutzbach,
1985; Kessler, 1985), the glacial evolution is
marked by a general ice retreat in the bolivian
Oriental Cordillera (Gouze et al., 1986), despite a

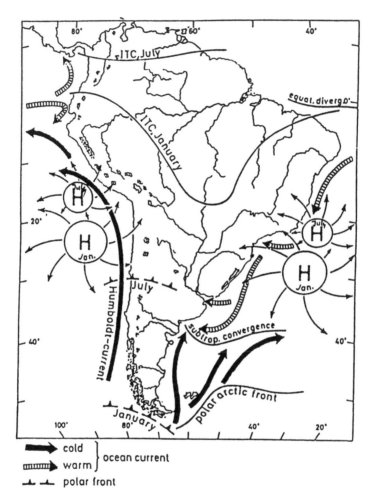

Figure 15. Interpretative curves of the Holocene oscillations for the lake Titicaca (this work), for the glacier (after: Gouze et al., 1986; Gouze, 1987) and for the peat-bog development (after: Graf, 1979 and Ostria, 1987 a-b).

temperature fall of 2oC at high altitudes from 4300 yr B.P. (Roux & Servant-Vildary, 1987). However, this change seems to be more related to a notable decrease of precipitation inducing a global change of water flowing, especially during the Early and Middle Holocene (Servant & Fontes, 1984); except two episodes of general erosion in the cordilleran valleys - between 7000 and 6000 yr B.P. and after 1500 yr B.P. - the environmental conditions were characterized by winter precipitations in association with a cooling trend. These climatic alterations hindered the development of peat-bog deposits in the cordilleran valleys (Graf, 1979; Ostria, 1987a and b) and caused the lowering of the water level (Pierre & Wirrmann, 1986). Parallel during the time intervals 7000 - 5000 and 3500 - 1400 yr B.P., the dense Amazonian forest, in the southeast of Bolivia, decayed and disappeared (Servant et al., 1981).

This review of the patterns of regional environmental modifications in Bolivia during the Holocene underlines that they are concomittant with the lake fluctuations, the glacier oscillations and the vegetation changes. According to the present knowledge, the main mechanism involved in these variations, is the modification of the mode of precipitation as inferred by the previous field evidence and by theoretical analysis on the water budget variations presented by Kessler (1963, 1985), Hastenrath (1981, 1985), Hastenrath and Kutzbach (1985), when the temperature fluctuations do not play a prominent role in these events.

ACKNOWLEDGEMENTS

We would like to thank Felix Alvarez, Charlie Barton and Ramón Catari for their helpful assistance in the field, J. Harle for her laboratory work, J. C. Fontes and M. Fournier who provided the 14C dating, V. Pittaluga for the correction of this English text, and Joyce who drafted the figures.

REFERENCES

Ahlfeld, F.E. 1972. Geología de Bolivia. 190pp. La Paz: "Los Amigos del Libro" Boulange, B.; Vargas, C. & Rodrigo, L.A. 1981. La sédimentation atuelle dans le lac Titicaca. **Rev. Hydrobiol. trop.**, 14(4):299-309.

Carmouze, J.P. & Aquize, E. 1981. La régulation hydrique du lac Titicaca et l'hydrologie de ses tributaires. **Rev. Hydrobiol. trop.**, 14(4):311-328.

Carmouze, J.P.; Arce, C. & Quintanilla, J. 1981. Régulation hydrochimique du lac Titicaca et l´hydrochimie de ses tributaires. **Rev. Hydrobiol. trop.**, 14(4):329-348.

Carmouze, J.P.; Arce, C. & Quintanilla, J. 1984. Le lac Titicaca: stratification physique et métabolisme associé. **Rev. Hydrobiol. trop.**, 17(1):3-11.

Collot, D.; Koriyama, F. & Garcia, E. 1983. Répartition, biomasses et production des macrophytes du lac Titicaca. **Rev. hydrobiol. trop.**, 16(3):241-262.

Gouze, Ph. 1987. La Cordillere Orientale de Bolivie: glaciations plio-pléistocenes. **Essai de paléohydrologie (30.000 ans B.P.- Actuel) d´aprés les oscillations des glaciers et la composition isotopique des macrorestes végétaux.** Thése 3o cycle, Univ. Paris-Sud, Orsay, 173p.

Gouze, Ph.; Argollo, J.; Saliege, J.F. & Servant, M. 1986. Interprétation paléoclimatique des oscillations des glaciers au cours des 20 derniers millénaires dans les régions tropicales; exemple des Andes boliviennes. C.R.Acad. Sc. Paris, sér. II, t. 303, 3:219-224.

Graf, K.F. 1979. Untersuchungen zur rezenten Pollen und Sporenflora in der nordlichen Zentralkordillere Boliviens und Versuch einer Auswertung von Profilen aus postglazialen Torfmooren. **Habilitation-schrift zur Erlangung der Venia** legendi, 104p., 24 pl.

Guerlesquin, M. 1981. Contribution a la connaissance des Characées d´Amérique du Sud (Bolivie, Equateur, Guyane Francaise). **Rev. Hydrobiol. trop.**, 14(4):381-404.

Haas, F. 1955. Mollusca: Gastropoda. Report No 17, Percy Sladen Trust Expedition to lake Titicaca in 1937. **Trans. Limn. Soc.** London, ser. 3, 1(3):275-308.

Hastenrath, S. 1981. **The glaciation of the Ecuadorian Andes.** A.A. Balkema(ed.), 159p.

Hastenrath, S. 1985. A review of pleistocene to holocene glacier variations in the tropics. **Zeitsch. fur Gletscherkunde und Glazialgeologie** Band 21, S.. 183-194.

Hastenrath, S. & Kutzbach, J. 1985. Late Pleistocene climate and water budget of the South American Altiplano. **Quat. Res.**, 24:249-256.

Iltis, A. 1987. Datos sobre las temperaturas, el pH, la conductibilidad eléctrica y la transparencia de las aguas de superficie del lago Titicaca boliviano (1985-1987). Convenio UMSA/ORSTOM, Informe N o 3 en español y francés, ORSTOM, 20p. La Paz.

Kessler, A. 1963. Uber Klima und Wasserhaushalt des Altiplano (Bolivien, Peru) wahrend des Hochstandes der letzten Vereinsung. Erdkunde, 17:165-173.

Kessler, A. 1985. Zur Rekonstruktion von spatglazialem Klima und Wasserhaushalt auf dem peruanisch-bolivianischen Altiplano. Zeitsch. fur Gletscherkunde und Glazialgeologie, Band 21, 107-114.

Kunzel, F. & Kessler, A. 1986. Investigation of level changes of lake Titicaca by maximum entropy spectral analysis. Arch. Met. Geoph. Biocl., Ser. B 36:219-227.

Lavenu, A. 1981. Origine et évolution tectonique du lac Titicaca. Rev. Hydrobiol. trop., 14(4):289-297.

Lavenu, A.; Fornari, M. & Sebrier, M. 1984. Existence de deux nouveaux épisodes lacustres dans l'Altiplano Péruano- bolivien. Cahiers O.R.S.T.O.M, sér. Géol., 14(1):103-114.

Lazzaro, X. 1981. Biomasses, peuplements phytoplanctoniques et production primaire du lac Titicaca. Rev. Hydrobiol. trop., 14(4):349-380.

Libermann Cruz, M. 1987. Impacto ambiental de un proyecto de irrigación en praderas naturales del Altiplano norte de Bolivia. in Memoria del 1er Congreso de Praderas Nativas de Bolivia, P.M.P.R., 21p. Oruro, Bolivia.

Mourguiart, Ph. 1987. Les Ostracodes lacustres de l'Altiplano bolivien. Le polymorphisme, son intéret dans les reconstitutions paléohydrologiques et paléoclimatiques de l' Holocene. These 3e cycle, Univ. Bordeaux I, 263p. + pl + annexes.

Mourguiart, Ph.; Carbonel, P.; Peypouquet, J.P.; Wirrmann, D. and Vargas, C. 1986. Late Quaternary paleohydrology of lake Huiñaymarca (Bolivia). Hydrobiologia, 143:191-197.

Oliveira Almeida, L.F. 1986. Estudio sedimentológico de testigos del lago Titicaca. Implicaciones paleoclimáticas. Tesis de Grado, Univ. Mayor de San Andrés, 136p. La Paz, Bolivia.

Ostria, C. 1987a. Végétation de haute altitude des Andes de Bolivie (exemple d´ une vallée glaciaire: Hichu Kkota, Cordillere Royale). p.25-29 in "Séminaire Paléolacs- Paléoclimats en Amérique latine et en Afrique (20.000 ans B.P.- Actuel)", v.1, ORSTOM, Bondy, p.29-30 janvier 87.

Ostria, C. 1987b. Phytoécologie et paléoécologie de la vallée altoandine de Hichu Kkota (Cordillere Orientale, Bolivie). These de l´Univ. P. et M. Curie, Paris 6, 180p. + annexes.

Pierre, J.F. & Wirrmann, D. 1986. Diatomées et sédiments holocenes du lac Khara Kkota - Bolivie. Géodynamique, 1(2):135-145.

Polunin, N.V.C. 1984. The decomposition of emergent macrophytes in fresh water. Advances in Ecol. Res., 14:115-165.

Quintanilla, J.; Calliconde, M. & Crespo, P. 1987. La quimica del lago Titicaca y su relación con el plancton. Oldepesca, Documento de Pesca n o 004, Convenio CAF/IMARPE/UMSA, 321p. Lima, Perú.

Richerson, P.; Widmer, C. & Kittel, T. 1977. The limnology of lake Titicaca (Perú-Bolivia), a large, high altitude tropical lake. Inst. Ecol., pub. no 14, Univ. of California. Davis, 78p.

Richerson, P.; Widmer, C.; Kittel, T. & Landa, A. 1975. A survey of the physical and chemical limnology of lake Titicaca. Verh. Intern. Verein. Limnol. 19:1498-1503.

Roux, M. & Servant-Vildary, S. 1987. Diatomées et milieux aquatiques de Bolivie. Application des méthodes statistiques a l´évaluation des paléotempératures et des paléosalinités. p.41-46 in "Seminaire Paléolacs-Paléoclimats en Amérique latine et en Afrique (20.000 ans B.P.-Actuel)", v. 1, ORSTOM, Bondy, 29-30 janvier 87.

Servant, M. 1977. Le cadre stratigraphique du Plio-Quaternaire de l´Altiplano des Andes tropicales en Bolivie. Supp. Bull. AFEQ, 1(50):323-327.

Servant, M. & Fontes, J.Ch. 1978. Les lacs quaternaires des hauts planteaux des Andes boliviennes. Premieres interprétations paléoclimatiques. Cahiers O.R.S.T.O.M, sér. Géol., 10(1):9- 23.

Servant, M. & Fontes, J. Ch. 1984. Les basses terrasses fluviatiles du Quaternaire récent

des Andes boliviennes.   Datations par le 14C.
Interprétation   paléoclimatique.   Cahiers
O.R.S.T.O.M, sér. Géol., 14(1):15-28.

Servant, M.; Fontes, J.Ch.; Argollo, J. & Saliege,
J.F. 1981.   Variations du régime et de la
nature des précipitations au   cours des 15
derniers millénaires dans les Andes de Bolivie.
C.R. Acad. Sc., t.292, sér. II, 17:1209-1212,
Paris.

Sheriff, F. 1979. Cartografía climática de región
andina   boliviana. Rev. Geográfica, 89:45-68
+ 2 mapas fuera texto.

Wirrmann, D. 1982. Primeros resultados sobre el
estudio de los   testigos del   lago
Huiñaymarca. Inf. dactil., 34p. La Paz,
Convenio UMSA/ORSTOM.

Wirrmann, D. 1987. El lago Titicaca: sedimentología
y   paleohidrología durante el   Holoceno
(10.000   años B.P.-   Actuel).   Convenio
UMSA/ORSTOM, Informe no 6, ORSTOM, 61p. +
anexos, La Paz.

Wirrmann, D. & Mourguiart, Ph. 1987. Oscillations
et   paléosalinities des lacs du Quaternaire
récent en Bolivie.   p.3-8 in "Seminaire
Paléolacs-Paléoclimats en Amérique   latine et
en Afrique (20.000 ans B.P.-Actuel)", v. 1,
ORSTOM, Bondy, 29-30 janvier 87.

Wirrmann, D. & de Oliveira Almeida, L.F. 1987. Low
Holocene level   (7700 to 3650 years ago) of
lake Titicaca (Bolivia).   Palaeogeogr.
Palaeoclim. Palaeoecol., 59:315-323.

PAULINA NABEL
CONICET-MACN, Buenos Aires, Argentina

6

Magnetic susceptibility as an expedite method for characterization and correlation of near shore sediments

ABSTRACT

The magnetic susceptibility studied on sediments from five shore cores of the proximal continental shelf area, south of Punta Médanos lighthouse (Buenos Aires province) allowed to use this information for their rapid identification. Definite zones with consistent behaviour of magnetic susceptibility, perfectly differentiated one from another have been obtained. The maximum values of magnetic susceptibility are related to those in ferric elements and minimum $CaCO_3$, whereas minimum values are related to higher ferrous contents and maximum $CaCO_3$. A close relationship between magnetic susceptibility values and mineralogical composition was observed. Aspects related to compositional and textural variations of the sediments are analized in relation with changes that took place in the sedimentary environment, such as paleoclimatic changes and sea level variations during the deposition. This simple and rapid technique appears as an useful tool to identify and correlate monotonous sedimentary sequences located near each other in the continental shelf environment, and as a secondary climatic indicator.

RESUMEN

El estudio de la susceptibilidad magnética de cinco testigos de la plataforma continental próxima, al sur del Faro de Punta Médanos, Prov. de Buenos Aires, ha permitido realizar una rápida identificación de zonas de susceptibilidad de comportamiento coherente, perfectamente diferenciables entre si.

Los valores máximos de susceptibilidad magnética están asociados a máximos relativos en el contenido de elementos férricos y mínimos de CO_3 Ca, mientras que los valores mínimos de susceptibilidad magnética están asociados a máximos relativos en el contenido de elementos ferrosos y de CO_3Ca. Se señala Ia dependencia entre los valores de susceptibilidad magnética y la composición mineralógica. Se analizan aspectos relacionados con las variaciones composicionales y texturales registradas en los sedimentos en relación a los cambios que tuvieron lugar en el paleoambiente de sedimentación, tales como cambios paleoclimáticos y variaciones en el nivel del mar.

Este método expeditivo es una herramienta útil en la identificación y correlación de secuencias sedimentarias monótonas, próximas entre sí, en ambiente de plataforma continental, y también puede ser utilizado como un indicador climático secundario.

INTRODUCTION

The knowledge of the succesive events that built up the shore line history arises, among others, from the study of near shore deposits under the present sea level, seeing that they conform a dynamic unity with nearby continental outcrops.

The purpose of this research is to analize the possibility of using magnetic susceptibility values of continental shelf sediments for their rapid identification, as magnetic susceptibility reflects mainly a property of the total sample mineralogy, to relate it to the sedimentary environment, and to analize the possibility of correlating cores extracted near each other.

Different authors, bearing in mind a geological approach in the use of this parameter, found that:
1) Magnetic susceptibility values from sediments is a sensitive indicator of stratigraphy. In Quaternary continental sequences, such as loess deposits and paleosols, is a valid criterion upon which to differentiate and characterize them (Jones and Beavers, 1964; Nabel and Spiegelman, in press, among others).

In lake sediments, the records of
paleomagnetic data obtained from different cores
are correlated mainly by their magnetic
susceptibility values using them as an
identification tool of the stratigraphic sequences
to form a single magnetogram of declination and
inclination against time (Creer et al., 1980;
Turner and Thompson, 1981).
2)   Magnetic susceptibility values were also
used as an identifier of facial changes. Distinct
sedimentary regimes such as "entrance" facies and
"interior" facies in cave sediments gave different
values in magnetic susceptibility, confirming that
this parameter contributes to a better
understanding of the cave environment
(Papamarinopoulos and Creer, 1983).
In the study of sea sediments, magnetic
susceptibility values were used in the
identification of the development of new
mineralogical specimens and for a better knowledge
of the deep sea environment (Harrison and Peterson,
1965; Lovlie et al., 1971; Ellwood, 1979).
In studies carried out in Australian lakes, a
striking resemblance between the water-depth curves
from lakes was verified, Bowler and Hamada (1971),
based essentially on variations of particle size
and carbonate chemistry, and initial magnetic
susceptibility curves (Barton et al., 1980)
suggesting that this parameter could be used as a
water-level indicator. Recently, (Liu et al.,
(1985)) used magnetic susceptibility values,
together with other chemical and mechanical
analysis as an identification tool of Quaternary
climatic changes.

LOCATION AND SAMPLING

Five cores were collected from Buenos Aires inner
shelf, south of Punta Médanos lighthouse, near
shore line, at lat. 36o56´S and long. 56o   40´W
(Figure 1). These cores are part of those collected
by the Servicio de Hidrografía Naval, during the
feasibility studies to construct a deep-water
harbour.
The cores have been named VP21, 3.8 m long;
VP22, 6.8 m long; VP4, 5.8 m long; VP51, 3.4 m long
and VP52, 9.5 m long, and they were extracted with
a vibracore system, from an area characterized by
channels with N-S direction at the bottom of the
sea (Parker and Violante, 1982)

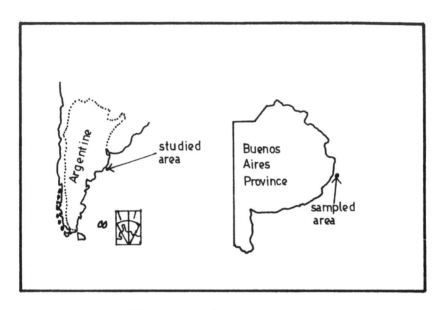

Figure 1. Location map.

between depths of 10.5 and 11m. These five cores
were extracted parallel to the SW-NE shore line
direction in this area, along a 1500 m distance.
The cores were split open and sub-sampled
systematically each 3 to 5 cm, obtaining 462
samples for paleomagnetic study. In all of them we
measured their initial magnetic susceptibility
values, which are presented in this paper. The
paleomagnetic study (Nabel, 1987) was started with
core VP21 and continued with VP22, but due to the
hydrodynamic characteristics of the environment,
the paleomagnetic results are not reliable enough,
particularly in the upper section of the sequence;
for this reason, the paleomagnetic study of the
rest of the cores was suspended. Nevertheless,
the consistent results obtained from the magnetic
susceptibility information encouraged us to present
them in this paper.

LITHOLOGY

Detailed macroscopic observation of the variations
in the structures, textures or colours of the

bottom sediments, allowed us to make some
generalizations from the lithological information
obtained.

Located to the north of the sampled area, the
cores, VP22, VP21 and VP4 present monotonous silt
argillaceous sequences with thin sand lenses and
caliche intercalations, distributed at the lower
section in VP22 and VP21 cores and lengthwise in
core VP4, all of them mainly dark yellowish brown
(10 YR 4/2) in colour. There are very small
quantities of shell debris, except at the top of
the cores, where all of them present approximately
50 cm of a silty shell-sand, olive gray (5 Y 4/1)
in colour and at the bottom of VP4 core, where
sandy sediments present isolated shells debris. At
different depths, the three cores present a
yellowish gray colour intercalation (5 Y 7/2)
within the same sediment texture (argillaceous
silt), with different thickness in each of the
cores. Core VP22 presents a thickness of 1 m,
whereas VP21, a thickness of 0.30 m and VP4, a
thickness of 0.70 m. This colour variation
coincides with a relative increase of the ferrous
contents in this level, in relation to the upper
sections of the profiles.

Table 1. Variation of ferrous/ferric contents in
relation with the changes in sediment colour.

Sediment colour	Fe_2O_3	FeO
Brown	3.22	0.40
Yellowish gray	3.12	0.66

The southernmost VP51 and VP52 cores are the
most sandy ones though the bottom of VP4 is also
composed of thin sand. The top of VP52 is composed
of argillaceous-silty sediment with caliche

intercalations, but from a depth of 1 m, there is a succession of sandy intercalations with shell debris and caliche, becoming more frequent and thicker at its base (10.64 m of depth).

VP51 core presents a silty-shell sand on the top, very similar and with the same colour of that one observed in VP22, VP21 and VP4. Below this upper section, an argillaceous silty sediment with caliche intercalations and sand lenses was observed; it becomes more frequent and thicker towards the bottom which is substantially composed of sand.

Particle-size determinations were also made and carbonatic reactions were obtained from each core. These results are later on presented in Figure 4.

STRATIGRAPHIC SITUATION

The studied sediments belong to two formations named Pozo No 10 Formation and Banco Punta Médanos Exterior Formation (Parker, 1980).

Pozo No 10 Formation, Pleistocene in age, occurs in marine and continental facies (Parker, 1979). The marine ones, characterized by pelitic sediments, intercalated with sand lenses and abundant shell debris, is presented at the base of cores VP4 and VP5 and in almost all (except the top) core VP52. This could probably be correlated with the "Belgranense" deposits (Frenguelli, 1957) or with the Pascua Formation (Fidalgo et al. 1973). The continental facies, characterized by the absence of marine fossils, their reddish colours and the high caliche in situ contents (Parker, 1979; Nabel, 1987) is presented in almost all cores VP21 and VP22 (except the top) and in the center of cores VP4 and VP51. It could probably be correlated with the "Bonaerense" deposits (Frenguelli, 1957) or with the Buenos Aires Formation (Riggi et al., 1986). The Banco Punta Médanos Exterior Formation is formed by transgressive sands of Holocene age with high quantities of shell debris (Parker, 1979). It could probably be correlated with the "Querandinense" (Frenguelli, 1957) or with the Destacamento Río Salado Formation (Fidalgo et al., 1973).

The paleomagnetic study from cores VP21 and VP22 gave a Bhrunes age (younger than 0.7 m.y.) for the sequence (Nabel, 1987).

In the Banco Punta Médanos Exterior Formation, radiocarbon ages of 10.380 +/- 180 and 11.610 +/- 140 yr B.P. were obtained (Parker and Violante, 1982). The Pozo No 10 Formation is too old for radiocarbon dating.

MAGNETIC SUSCEPTIBILITY

Bulk initial magnetic susceptibility was measured from 462 samples belonging to the five cores. Smoothed values against depth for the sediments at the bottom, can be observed in Figure 2. In each core it was possible to separate sections where the magnetic susceptibility behaviour is consistent and different in their mean values from the upper or lower zones.

The VP22 (6.80 m long) susceptibility profile presents, for the 50 cm shell-sand on the top, susceptibility mean values of 13 u6/Oe; this section of the profile was named Zone I. After screening the coarser fraction of the sediments of this zone, consisting mainly of shell debris, we found susceptibility values of 120 u6/Oe in the remaining sandy silt sediments, of the same magnitude as the inferior zone. This second Zone, constituted by brown argillaceous silt, that was continued up to 2.70 m of depth, was called Zone II. Towards the bottom and without any variation of macroscopic behaviour like lithological or structural changes, and up to 5.80 m of depth, it is possible to recognize a section where susceptibility mean values fluctuate above 50 u6/Oe and was named Zone III. Under this section it is possible to distinguish the colour change in the pelitic sediment mentioned, which is coincident with a new decrease in the susceptibility mean values to 12 u6/Oe, this section continues up to the bottom of the core and it was named Zone IV.

The susceptibility profile of core VP21 (3.80 m long) presents the same susceptibility behaviour in the first three sections, but with different thickness in each of them, named Zone I, II and III. At the bottom of this core, it was observed a change in the sediment colour and a relative decrease in the susceptibility values but this behaviour was not clear enough from the susceptibility point of view, to undoubtedly separate a susceptibility zone.

135

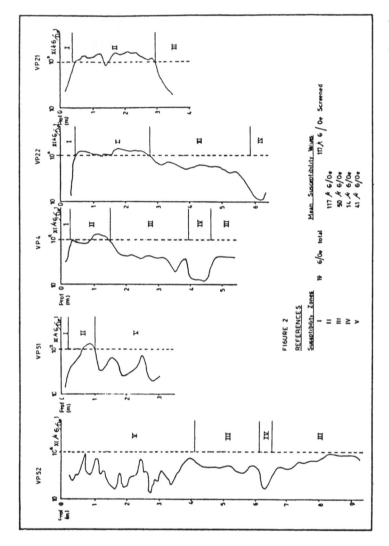

Figure 2. Smoothed magnetic susceptibility logs vs. depth, showing delimitated zones which were marked with roman numbers. Curves were plotted on logarithmic scales.

The susceptibility profile of core VP4 (5.80 m long) presents the same succession of susceptibility zones described for core VP22 but with a different thickness, which are recognized within a similar lithology. They were also named Zone I, II, III and IV. An increase of the susceptibility values was also observed with a similar behaviour to that defined for the upper Zone III under the last yellowish gray pelitic level, which presents minima susceptibility values, and from 4.60 m of depth up to the base of the core.

The susceptibility profile of core VP51 (3.40 m long) presents at the top and coincident with the 45 cm of shell-sand sediment an equivalent susceptibility behaviour as those defined for Zone I in cores VP22, VP21 and VP4. Under this level, a thin section corresponding to Zone II was observed. Below this and up to the base of the core, the susceptibility values are characterized by succesive maxima and minima. Although the mean values are close to 50 u6/Oe, the peak design of this section makes it different from the others. It was named Zone V.

In its upper section, the susceptibility profile of core VP52 (9.40 m long) presents a similar behaviour to that of Zone V defined in core VP51; meanwhile, in the lower section from a depth of 5.30 m, it presents a similar susceptibility behaviour to that characterized for Zone III in cores VP22, VP21 and VP4. Between 7.50 and 8.10 m of depth, this section presents an intercalation of lower values in susceptibility of the same order that those from Zone IV in VP22 and VP4 cores. In this case, the minima values are not coincident with any variation in sediment colour or some other lithological or structural change macroscopically evident.

DISCUSSION AND CONCLUSIONS

The changes in the magnetic susceptibility behaviour of the different cores, as they are reflected by zones succession in each of the profiles (Figure 2), make evident that there is a systematic variation of this parameter even though macroscopical facial or compositional changes were not observed. Meanwhile, when macroscopically changes such as variations in the sediment colour

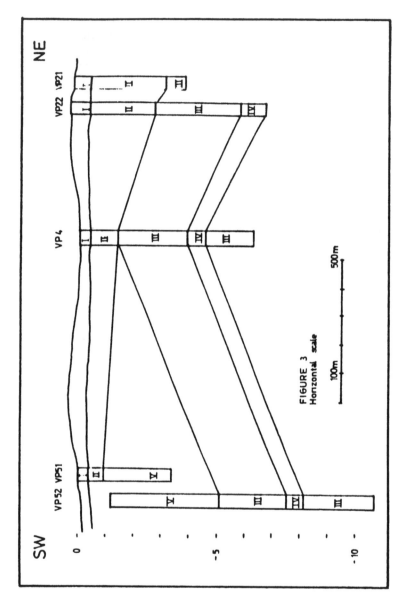

Figure 3. Schematic profile with NE-SW direction obtained from the correlation of the magnetic susceptibility logs.

or higher shell debris contents were observed, they
are always reflected by variations in the
susceptibility values, indicating that magnetic
susceptibility is a sensitive method to recognize
changes in sedimentary records.

Available information concerning the coring
procedure indicates that core VP22 would have been
taken from underneath core VP21 and would represent
only the lower section of the sedimentary column.
The same would have happened with core VP52 related
to core VP51. The study of the magnetic
susceptibility allowed us to discuss the validity
of such data. The results obtained from core VP22
show that it is a complete and independent core,
whereas core VP52 shows a lack of the upper part of
the sedimentary column, which was probably washed
out from the core during the coring operation.

The effect of particle size on susceptibility,
as it is reflected by the presence of ferrimagnetic
minerals is thought to be negligible.

Cores VP51 and VP52 mainly present a sandy
composition whereas cores VP21, VP22 and VP4 are
more argillaceous or silty in texture though all of
them have not a pure texture but a mixture of them.
Probably as a consequence of it, the magnetic
susceptibility results from the five cores are
similar and comparable among them. Although the
screening of the coarser section from Zone I caused
an important change in susceptibility values, this
happened because of the diamagnetic contribution of
the shell debris located in that fraction. The
variations observed in magnetic susceptibility
values are interpreted as being caused by changes
in mineralogical contents. Maximum values in
magnetic susceptibility are coincident with
relative maximum of ferric oxides in the sediments
(Zone II); meanwhile, minimum values are related
with higher proportions of ferrous oxides (Zone IV)
or with higher proportions of carbonates as in Zone
I with shell debris, or Zone V, where the peak
design reflects weak paramagnetic and diamagnetic
minerals such as caliche minerals mixed with the
ferrimagnetic ones.

The defined susceptibility zones from each
core can be compared and correlated as shown in
Figure 3. Taken into account that the cores were
extracted from a similar depth under sea level,
in the same topographic environment, the tops of
the profiles VP22, VP21, VP4 and VP51 were flushed.

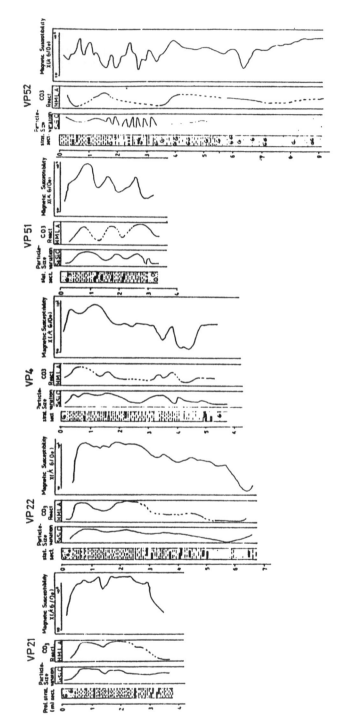

Figure 4. Comparison between initial magnetic susceptibility curves with stratigraphic sections, particle-size variation and carbonatic reaction curves.

To define the location of core VP52 the magnetic susceptibility and coring information were used. This correlation makes evident that it is possible to perform a compound profile from magnetic susceptibility behaviour, parallel to the shore line, indicating a susceptibility zonation, which reflects environment changes, and would be used as an indicator of the extent of these changes.

The determination of particle-size variations and carbonatic contents of the studied near-shore sediment, was plotted against depth in Figure 4. In this figure, it is possible to compare the behaviour from these parameters with the initial magnetic susceptibility curve of each core. That comparison points out the high similarity among them, confirming magnetic susceptibility sensitivity to record environment changes. However, this parameter could not be used isolately. Magnetic susceptibility as well as grain size distribution, $CaCO_3$ content and other geochemical indicators are tools to obtain valuable information on the paleoenvironment (Liu et al., 1985) but, in order to use them, it is necessary to hold a detailed geological description of the depositional environment so as to analize correctly the changes recorded by these parameters.

Relative maximum magnetic susceptibility values, presented in Figure 4, are coincident with relative minima in $CaCO_3$ content, in grain size and, although it is not shown, with minima in ferrous contents. This behaviour could reflect a higher water level in sea or lake environments as well as continental loessic sequences.

Relative minima in magnetic susceptibility values, coincident with relative maximum in $CaCO_3$, ferrous and grain size contents could reflect sea or lake shore environments, restricted shores or caliche deposits in a continental environment.

The common denominator of these changes is the climate.
Climatic fluctuations serve to change sea level by the growth of continental ice caps, the chemistry and behaviour of the oceans and the salinity of lakes (Fisher, 1982) as well as determinant of the presence of loess, paleosol or caliche deposits in continental environment.

Taking into account that the studied near-shore sediments include continental as well as marine environments, it is necessary to recognize them in the cores in order to avoid mistakes and make appropiate comparisons.

The continental loessic sections of the cores VP21, VP22, VP4 and VP51 which present relative minimum susceptibility values, coincident with higher $CaCO_3$, grain size and ferrous contents, could be used as indicator of cold and dry conditions.

In general, the coarser grain size implies a larger wind speed and a drier climate. A higher $CaCO_3$ content is also characteristic of soil and loess in dry and cold environments as compared with those of humid and warm environments where the soluble $CaCO_3$ is usually removed by percolating water. In dry and cold climates, iron exists as FeO, but in a humid and warm climate Fe_2O_3 is the prevailing state (Liu et al, 1985). According to this viewpoint, relative maximum of magnetic susceptibility values in our continental sections, could be interpreted as related to warmer and more humid conditions.

The relative minimum values of magnetic susceptibility of the marine environment in sections presented at the top of cores VP21 and VP22, at the top and at the base of cores VP4 and VP51 and in core VP52, coincident with relative maximum from the other parameters, reflect a shore environment, representing relative colder conditions. On the contrary, relative maximum magnetic susceptibility values in this environment (coincident with minimum values in the other parameters), represent deeper conditions, i.e. warmer climates. In both, continental and marine environments, higher magnetic susceptibility values are coincident with warmer climates, while the lower values are coincident with colder climates. This coherence between magnetic susceptibility behaviour and climatic changes suggests the possibility to use this parameter as a secondary climatic indicator.

Nevertheless, it is necessary to analize the shore situation as a whole to make accurate interpretations of the magnetic susceptibility meanings obtained from each environment.

In a first approach, it is possible to suggest that in the loessic section of cores VP21, VP22 and VP4, the smoothed susceptibility values exhibit a general trend to a climatic cycle relatively

longer which began with a dry and cold climate
(base of core VP22) and became more humid and
warmer throughout time, finishing with a marine
transgression (top of core VP22). Cores VP21 and
VP4 showed the same cycle. In all of them lower
variations in susceptibility values could reflect
changes in a lower scale.

In core VP51 three cycles with a smaller wave
length that could reflect changes in a lower scale
were observed, but taking into account its wider
trend towards the top, it could be probably
included in the larger cycle exhibited by the other
cores.

In core VP52, deposited almost completely in a
marine environment, the maximum and minimum
susceptibility values probably reflect relative
higher and lower sea levels, related to a lower
scale of climatic changes.

The data reported above indicate that
magnetic susceptibility is an easy and rapid
technique to use as an aid in characterizing and
correlating sedimentary sequences located close to
each other, in continental shelf environment and
could be also used as a tool to estimate
paleoclimatic changes to a certain extent.

ACKNOWLEDGEMENTS

I am grateful to CONICET that makes this work
possible, to the Argentine Servicio de Hidrografia
Naval, that provide the study material, to Dr. G.
Parker for his generous disposition, to the last
Ing. D. Valencio for the use of the paleomagnetic
laboratory and to Dr. L. Spalletti for useful
suggestions.

REFERENCES

Barton, C.E. 1983. Results from Australia.
  Geomagnetism of backed clays and recent
  sediments. Creer, K.; Tucholka, R. & Barton,
  C. (eds.). p.236-243.
Bowler, J.M. and Hamada, T. 1971. Late Quaternary
  stratigraphy and radiocarbon chronology of
  water level fluctuations in lake Keilambete,
  Victoria. Nature, 232:330-332.
Creer, K.M.; Hogg, T.E.; Rezman, P.W. & Reynaud, C.
  1980. Paleomagnetic secular variation curves

extending back to 13400 years BP recorded by sediments deposited in Lac de Joux, Switzerland. J. Geophys., 48:139-147.

Ellwood, B.B. 1979. Sample shape and magnetic grain size: two possible controls on the anisotropy of magnetic susceptibility variability in deep-sea sediments. Earth and Pl. Sc. Letters, 43:309-314.

Fidalgo, F.; Colado, U.& De Francesco, F. 1973. Sobre ingresiones marinas cuaternarias en los partidos de Castelli, Chascomús y Magdalena (Pcia de Buenos Aires). Actas V Congreso Geológico Argentino, IV:27-39.

Fisher, A.G. 1982. Long-term climatic oscillations recorded in stratigraphy. Climate in Earth History. National Academy Press:97-104.

Frenguelli, J. 1957. Neozoico. Geografía de la República Argentina. Gaea, 2:1-115.

Harrison, C.G.A. & Peterson, M.N.A. 1965. A magnetic mineral from the Indian Ocean. The American Mineralogist, 50:704-712.

Jones, R.L. & Beavers, A.H. 1964. Magnetic susceptibility as an aid in characterization and differentiation of loess. J. Geoph. Res., 69:881-883.

Liu, T.; Zhisheng, A.; Boayin, Y. & Jimao, H. 1985. The Loess- Paleosol sequence in China and Climatic History. Nature, 8(1):21-28.

Lovlie, R.; Lowrie, W. & Jacobs, M. 1971. Magnetic properties and mineralogy of four deep-sea cores. Earth and Pl. Sc.Letters, 15:157-168.

Nabel, P.E. 1987. Estudio paleomagnético y sedimentológico de sedimentos de plataforma. Rev. Asoc. Geol. Arg., VLII.

Nabel, P.E. & Spiegelman, A.T. (in press). Caracterización sedimentológica y paleomagnética de una sección del Pampeano, en el subsuelo de la Ciudad de Buenos Aires. Rev. Asoc. Geol. Arg., Buenos Aires.

Papamarinopoulos, S. & Creer, K.M. 1983. The paleomagnetism of cave sediments. Geomagnetism of backed clays and recent sediments. Creer, K.; Tucholka, R. & Barton, C. (eds.). 243-249.

Parker, G. 1979. Geología de la planicie costera entre Pinamar y Mar de Ajó, Prov. de Bs. As. Rev. Asoc. Geol. Arg., XXXIV (3):167-183.

Parker, G. 1980. Estratigrafía y evolución morfológica durante el Holoceno en Punta Médanos (Planicie costera y plataforma interior), Prov. de Bs. As. Simposio sobre problemas geológicos del litoral atlántico bonaerense. Resúmenes. CIC:205-224. La Plata.

Parker, G. & Violante, R.A. 1982. Geología del frente de costa y plataforma interior entre Pinamar y Mar de Ajó, Prov. de Bs. As. Acta Oceanográfica Argentina, 3(1):57-91.

Riggi, J.C.; Fidalgo, F.; Martinez, O. & Porro, N. 1986. Geología de los "Sedimentos Pampeanos" en el partido de La Plata. Rev. Asoc. Geol. Arg., XLI (3-4):316-333.

Turner, G.M. & Thompson, R. 1981. Lake sediment record of geomagnetic secular variation in Britain during Holocene Time. Geophys. J. Rr. Astr. Soc., 65:703-725.

CLAUDIO A.SYLWAN
Geological Institute, University of Stockholm, Stockholm, Sweden

7

Patagonian Pleistocene glacial varves:
An analysis using variation of their thickness

ABSTRACT

Six sequences of glacial varves from Patagonia have been analyzed. Spectral power analysis was performed on the total varve thickness in all six sequences, and on individual thickness of summer and winter units in two of the sequences. The 11- and 22-years cycles appear conspicuosly indicating the influence of the corresponding solar cycles. A 18.2-years cycle has been recorded in one sequence representing the last deglaciation period. This suggests the influence of the 18.6-years lunar tidal cycle. A sequence representing maximum glacial conditions reveals a strong 8.5-/8.8-years cycle.

RESUMEN

Seis series de varves glacilacustres de la Patagonia han sido analizadas con respecto a la variación del espesor de cada unidad anual. El método de Fourier de análisis espectral fue aplicado sobre la variación del espesor total de los varves de las seis series, y sobre la variación del espesor de los términos estacionales de invierno y verano de dos de aquellas series. Ciclos de 11 y 22 años se observan claramente, por lo que se infiere el efecto de la actividad solar. Una ciclicidad de 18,2 años se registra en una serie de varves depositada durante la última deglaciación, lo que indicaría los efectos del ciclo lunar de mareas de 18,6 años. Una serie várvica representando un período de máxima glaciación, registra una fuerte ciclicidad de 8,5 - 8,8 años.

INTRODUCTION

By definition "a varve (Swedish: "varv", layer) is any sedimentary bed or lamination deposited within the period of one year, or any pair of contrasting laminae representing seasonal sedimentation (as summer and winter) within the period of one year" (Morner, 1978).

The study of varves began in 1884 when De Geer (1940), undertook the first varve measurements in the Stockholm area. In 1904, De Geer started a real "varve campaign" and in 1910, he was able to present a detailed and well dated picture of the ice recession after the last glaciation maximum (De Geer, 1912) at the International Geological Congress in Stockholm. A new geochronological tool was established. As a result of this, the method was adopted by several researchers who applied it in other regions. In southern Finland, Sauramo (1923) carried out a very detailed study on the varved sediments. De Geer's students were encharged to test the so-called "teleconnection" between distant zones. Hence, Norin (1927) was sent to the Himalaya area, Nilsson (1932) to eastern Africa, Antevs (1922) to North America in order to study the recession of the ice sheet in New England and Caldenius (1932a, 1932b) to Patagonia and Tierra del Fuego (he also visited New Zealand). The masterly work that Caldenius carried out in southern Argentina was unfortunately "teleconnected" with the Swedish geochronologic scale of De Geer (1940). This kind of "long distance correlations" have many times been criticized and objected to. Recent paleomagnetic analysis carried out in the Lago Buenos Aires Valley, Patagonia, conclusively showed that this correlation is no longer tenable and consequently that the terminology and proposed ages must be revised (Morner and Sylwan, 1987, 1988).

Many attempts to prove the annual nature of the varved sediments have been done (Hogbom, 1889; Johnston, 1922; Terasmae, 1963), achieving most of them satisfactory results. Among the evidence that support the yearly cyclicity of the varves, there are some of conclusive character.

Arrhenius (1947) made important contributions to the knowledge of the origin of colour varvity in glacial clay, in investigations carried out on varved marls in Upsala. He analyzed the disolution of the winter material of marl varves by hydrocarbonic acid, which in winter is much greater than in summer.

Kuenen (1951) discussed the theoretical mechanisms of varve formation and proposed a model (the so-called De Geer-Kuenen model) for glacial varve deposition. According to this model, the coarser units are formed by turbidity currents originating from ice melting in the summer (without producing erosion in the underlying layers). Towards the top, the summer lamina grades into the winter lamina, which is composed of the suspended matter that was brought into the lake basin during summer, but, because of its small grain sized sediments during winter, when calm conditions predominate. Summarizing, three genetically different parts of a varve can be identified:

1. Summer unit deposited due to the action of turbidity currents.
2. The winter unit comprises two parts: 2a deposited by turbidity currents after stagnation, and 2b deposited by slow, continuous settling from suspension.

It is assumed that part 2a is insignificant and that the bulk of winter unit consists of part 2b.

Agterberg and Banerjee (1969) established a semi-quantitative model for varve deposition. According to this model the individual varve thickness is mainly controlled by:

1. Rate of retreat of the ice front at the time of deposition (controls the amount of meltwater and the amount of sediment supply).
2. The distance of the formation place to the ice front (determined by the rate of ice retreat prior to time).
3. The variation of thickness as a function of distance from the ice front.

Recent studies comparing varves from the Precambrian up to the Holocene (Anderson & Dean, 1988), indicate that glaciolacustrine varves have retained their distinctive character through time, which suggests that the mechanisms of glacial varve deposition have not changed since Precambrian times.

De Geer (1940) made a detailed study of the pattern of variation of the thickness of the varves and its relationship with the position of the ice front. According to this he defined four different facies (from the bottom: 1- ose-centre or submarginal delta, 2- extramarginal current-ridge of stone-free sand, gradually passing to 3-ordinary clay-varve, proximal and distal which ultimately has a 4- very considerable microdistal extension).

Because varve thickness is a function of ice melting, variations of its thickness should reflect variations in certain climatic parameters, e.g. temperature, precipitation. Granar (1956) found in northern Sweden a good correlation between postglacial varve thickness and annual stream discharge and Stenborg (1970) that ablation is sensitive to temperature and precipitation. Likewise, Perkins and Sims (1983) obtained certain correlation between mean annual temperature and variation in varve thickness in a short sequence of Alaskan varves. Concerning the sun radiation, it is known that 11- and/or 22-years cycles have been recognized in varve data from Recent Pleistocene and pre-Pleistocene times (Anderson, 1961; Sonnet and Williams, 1985; Williams and Sonnet, 1985). Moreover, variations in high-water levels in Finnish lakes have been found to follow the periods of solar activity (11 and 22-23 years) and with obvious similarities the periods of lunar circulation (18 and 31 years; Keranen, 1984).

Six varve sequences from Patagonia will be analyzed. Spectral power using the Fast Fourier Transform (FFT; Menke, 1986) will be applied in order to examine if there are any particular cyclicity in the variation of the varve thickness. In all the FFT spectral power analyses here presented, a trend computing a least-square fit between number of varve and varve thickness is employed (Menke, 1986), appart from this treatment no other filter or smoothing was used on the raw data.

VARVE SEQUENCES IN PATAGONIA

Caldenius (1932a, 1932b) made an outstanding work not only in varve-geochronology but also he masterly mapped the four zones of maximal ice extensions in southernmost South America (considered very exact even today). By field observations in the whole of Patagonia he was able to distinguish four different glaciations, or more exact three moraine systems and an older one that might represent several glaciations. Concerning the geochronological work, Caldenius measured in detail

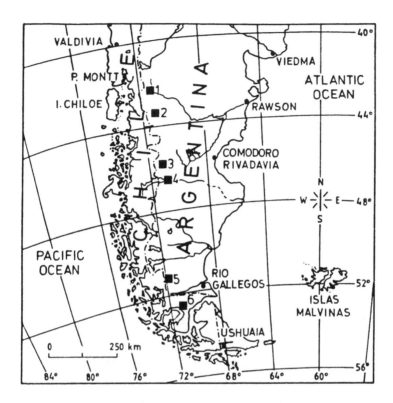

Fig. 1. Geographical setting of the Six localities
where varve sequences have been taken. (1) Lago
Epuyén, (2) Río Corintos, (3) Lago Blanco, (4) Lago
Buenos Aires, (5) Río Vizcachas, and (6) Laguna
Blanca.

several varve series in four different main
localities.

Data from six localities (five in Argentina
and one in Chile), have been taken from the
existing literature and from our measurements, as
follows:
1. Lago Epuyén (Caldenius,1932a), 42o10´S,71o30´W.
2. Río Corintos (Caldenius,1932a), 43o10´S,71o20´W.
3. Lago Blanco (Sylwan,1988), 45o50´S,71o10´W.
4. Lago Buenos Aires (Sylwan,1987), 46o35´S,71oW.
5. Río Vizcachas (Caldenius,1932a), 51oS,72oW.
6. Laguna Blanca (Caldenius, 1932a),52o30´S,71o10´W.
The geographical setting of the localities is
shown in Figure 1.

The data from Caldenius' localities have been, in this paper, redrawn without using the number he assigned to every varve, instead we assigned number one to the lowermost varve. We do this because the numbers used by Caldenius are related to the Swedish geochronologic scale of De Geer (1940), which, as already said, is in no way tenable for the Patagonian data. All the curves of thickness of the varves are accompanied by small inserts showing the cumulative curves of each deposit (absissae representing years and ordinates the size of the deposit). The measurements published by Caldenius (1932a) do not include the separate thicknesses of the winter and summer units; for this reason, Lago Buenos Aires and Lago Blanco will be treated first.

LAGO BUENOS AIRES

This area was first investigated by Caldenius (1932a), who mapped the moraine-systems and the glaciolacustrine sediments. A sequence of more than 740 glacial varves representing the ice recession was measured. By applying the so-called teleconnection he assigned to the bottom varves of the sequence an age of 10,000 years B.P.
    In the valley of Lago Buenos Aires, 887 late glacial varves were counted and the thickness of each single varve (Figure 2a) as well as the thickness of the summer and winter units individually were measured (Figure 2b) in detail. The age of the entire succession ranges between 13,500 and 12,500 years B.P. (age obtained from two radiocarbon datings; Sylwan, 1987). This sequence is obtained by connecting two profiles ("A" and "B"). Unfortunately, these two profiles have no level in common, however. Because of the short distance between them and their stratigraphical position, it is assumed that only two or eventually three basal varves could be missing at around varve 60 (in this bottom section the mean varve thickness is around 10-15 cm). The decrease in thickness with time and the gradual gradation from coarser grain size of basal varves to finer grain size in top varves are evidences for a simple and single ice recesssion (De Geer, 1940; Ignatius, 1958). At the bottom, the winter units represent, at a mean, 10-15% of the total varve thickness changing gradually upwards to 40-50% at the top (Figure 2c shows this relationship between varve 1 and 610). Ashley

(1975), explains such a change in terms of ice recession. The rate of ice recession at Lago Buenos Aires was calculated (Sylwan, 1987) to be about 160-165 m/year.

Figure 2b shows the absolute thickness of summer and winter units. There is a good correspondence between extra thick summer units and extra thick winter units. This feature illustrates the relationship between the thickness of summer and winter units when abundant material has been brought into the basin during summer.

Although the total varve thickness experiences a general and marked decrease with time, there are some peaks, around varves 100-130, 260-300 and around varves 410-420, which may actually be interpreted as small oscillations in the retreat of the ice. Agterberg and Banerjee (1969) interpreted a similar behaviour as typical of "surging" glaciers (see Flint, 1971).

FFT spectral power on the total thickness of the varves was applied only to the lowermost 550 varves of the main sequence (profile "B"). This selection was done in order to avoid the influence of the uppermost section of the series, consisting of microvarves with little or no measurable variations in thickness, and for avoiding the gap of missing varves between the two profiles. Time series analysis on total varve thickness (Figure 3a) shows major density of power for periods of around 22.7, 19.3 years, a concentration around 11 years (peaks at 12.3, 11.3 and 10.1 years), another concentration around 7 years (peaks at 7.0 and 7.2 years), 5.9 and 3 years. A long term period of about 140 years may actually reflect the above mentioned small oscillations in the ice retreat.

Due to the different genesis of the deposition of summer and winter units, FFT spectral power analysis was also applied to both these units separately.

The spectral analysis on data of summer units (Figure 3b) shows concentration of peaks at 22.7, three peaks at around 11 years, two peaks at around 7 years and a peak of density at around 3 years. This pattern, in general lines, agrees well with the one obtained from the spectrum of the total varve thickness, these similarities were actually expected. On the contrary, the spectrum obtained from the winter data (Figure 3c) differs markedly. The most important peaks of power are the 18.2- and 11.3- years cycles.

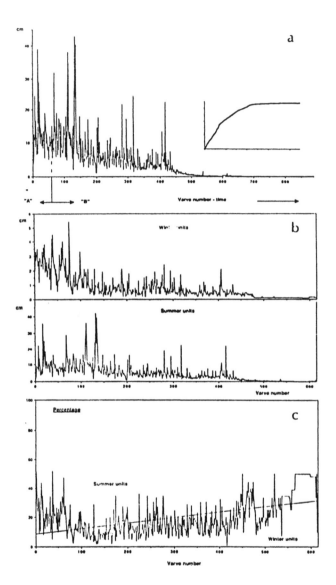

Fig. 2. Data from Lago Buenos Aires. (a) Thic kness
of the combined varve sequence versus time. "A" and
"B" indicate the two different profiles measured,
the insert represents the cumulative curve of the
growth of the varves. (b) thickness of the winter
units - up - and the summer units - down - versus
time. (c) Percentage of the winter units with
respect to summer units, the line indicates the
best fitted slop for this variation.

154

The obtained results are very interesting. Both 11- and 22- years cycles are present in the summer data. The 11- and 22- years solar cycles have many times been recorded in varves (Schove, 1983; Sonnet and Williams, 1985), while the 3-years cycle has commonly been described (Schove, 1979). The 18.6-years cycle, attributed to the lunar tidal cycle (Pettersson, 1912; Fairbridge, 1984; Fairbridge and Sanders, 1987), has been found in several South American records, viz. from Punta Arenas and Santiago de Chile in records of temperature variation (Currie, 1987), and from Buenos Aires, Santiago de Chile and Rio de Janeiro in records of pressure variation (Currie, 1987). In a tree-ring record from the Patagonian Andes, the 18.6-years cycle was found together with the 11-years solar component (Currie, 1983). Although we have no evidence yet supporting that the varves from Lago Buenos Aires are recording variations due to these lunar tidal and solar components, the results are suprisingly similar. However, the relationships found by Keranen (1984) between variations in high-water level in Finnish lakes with lunar tidal cycles, led to a better understanding of the phenomenon in Lago Buenos Aires. It, therefore, seems as if, for first time, are recording such a lunar tidal periodicity in late Pleistocene varves.

LAGO BLANCO

In the Lago Blanco basin, 178 glacial varves, interpreted as formed in close association with the last glacial maximum (15,000?, 20,000? years B.P.; Sylwan, 1988), have been measured in detail. Variation of the thickness of the varves and their seasonal terms are shown in Figure 4a, from which a general increase in the thickness of varves is observed. Around varves 70-80 and increase in the mean rate of sedimentation is interpreted in terms of the culmination and stagnation of an advancing ice margin.

Figure 4b shows the variation in thickness of both summer and winter units versus time. The main characteristic of the varves of the curves is the marked periodicity in the variation of the thickness, which can even be appreciated by eye. Some conspicuous "reversals" (winter units larger than the summer ones in a single varve) are shown in Figure 4b.

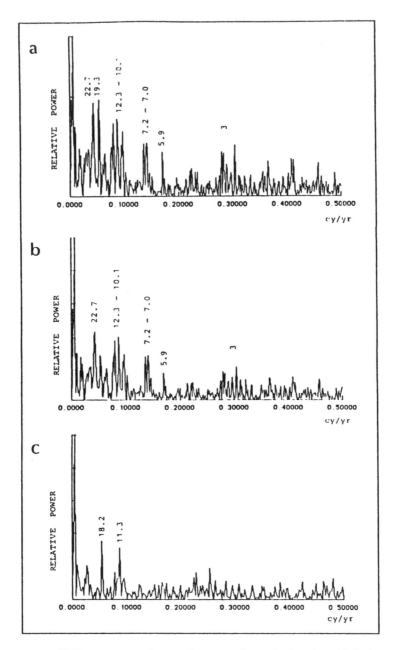

Fig. 3. FFT spectral analyses for (a) the thickness of the varves, (b) the thickness of the summer units, and (c) the thickness of winter units from Lago Buenos Aires.

FFT spectral analyses have been applied to the data obtained from the varves from Lago Blanco (Figure 5). The major and single well defined peak of power density of the spectrum of total varve thickness (Figure 5a) is 8.8 years. A second peak of 85 years is identified. Because the data set covers only 178 years, little or no reliability is placed on this, however. Less significant peaks are 43-, 26-, and 5- years cycles. The results obtained from summer units data (Figure 5b) agree almost totally with the latter; only the 5-years cycle is missing. The major power density from winter units data (Figure 5c) is represented by a 8.5 years cycle. Two minor peaks, 5.6- and 4.5-years, which probably represent the 5-years peak in the analysis of the total thickness, are also recorded.

A delay, generally of two years and sometimes of one year, in the peaks of the summer units maxima with respect to corresponding peaks of winter units is observed (Figure 4b). This is a very confusing result concerning the mechanisms of formation of the varves. According to the De Geer´s (1940) and Kuenen´s (1951) model for varve formation, the summer unit deposits by turbidity currents produced by melt water during summer, while the dominant amount of winter unit slowly settles down when calm conditions predominate.

The observed 1-, 2-years delay of summer peaks with respect to winter peaks (Figure 4b) can not be explained with the De Geer-Kuenen model, rather, other factors should also have influenced. The difficulty in explaining how a winter unit can be wider, and sometimes much wider, than the summer unit laid down the previously season, lies on the fact that the model considers only the summer as the period when the major amount of material enters the basin. It is worthy to point out that the main peak obtained from the winter time series analysis is a 8.5-years cycle, and the corresponding one from the summer data is 8.8-years cycle, this difference is consistent with the 1-, 2-years delay described above. Agterberg and Banerjee (1969) found a similar 2-years delay in varves from northern Ontario and a similar out of phase related to the 11-years solar cycle. They attempted to explain the phenomenon as being produced by a sudden increase of meltwater which produces very thick winter units ("reversals"), and the response of the summer unit delayed because a denser turbidity current dumped more material at the proximal end, leaving the distal varves unaffected.

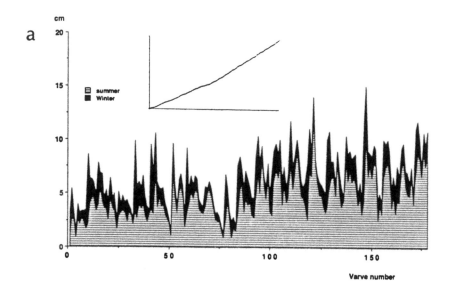

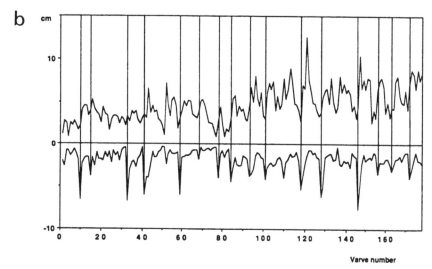

Fig. 4. Data from Lago Blanco, (a) thickness of the varves detailing seasonal units, insert represents the cumulative curve for the growth of the varve thickness, (b) curves showing the summer units - up - and the winter units - down - versus time. The vertical lines indicate peaks in winter units, the peaks of summer units occurring generally 2 years but also 1 year later.

But later the steeper profile of the underlying varve acting as a floor to the turbidity currents for later years helped them to spread further with the thicker part of the profile moving to more distal areas.

RIO CORINTOS

This zone has been excellently investigated by Caldenius (1932a). He mapped the different moraine-systems and measured the glacial varves which were interpreted as simultaneous with those from Lago Buenos Aires. Caldenius found that this sequence should represent a period of 586 years (amount of varves graphically represented in Caldenius, 1932a, plate 17) admiting that some (21) varves were missing (around varve 50; Figure 6a). Unfortunately, he did not published the values of the seasonal deposit thicknesses, which would have been good in terms of analyzing the series in the way done with Lago Buenos Aires and Lago Blanco.
        Caldenius (1932a) found three maxima in the curve of varve thickness and interpreted them as follows, the first was formed because the proximity of the ice margin, the second because of "climatological reasons", and the third because of the proximity to the lake shore at that moment.
        FFT spectral analysis of the varve thicknesses from the Rio Corintos sequence were performed using the values of the 500 uppermost varves, in order to avoid the zone where some varves are missing. The obtained spectrum (Figure 6b) shows main peaks of density at 51-, 30.1-, 21-years, a concentration between 11- and 10.2-years, 7.4-, 5- and 3-years cycles.
        There are many similarities between this spectrum and the one from Lago Buenos Aires; peaks at 3-, 5- and 7-years cycles are present in both sequences, peak around 10.2-11 years in Rio Corintos sequence and high density of power at 10-12-years at Lago Buenos Aires, peak at 21 years in Corintos and high density in 19-22 years in Lago Buenos Aires. These similarities seem to support Caldenius' (1932a) conclusions about a contemporaneous deglaciation in both basins. The 18-years term recorded in Lago Buenos Aires is absent in the record from Rio Corintos, however.

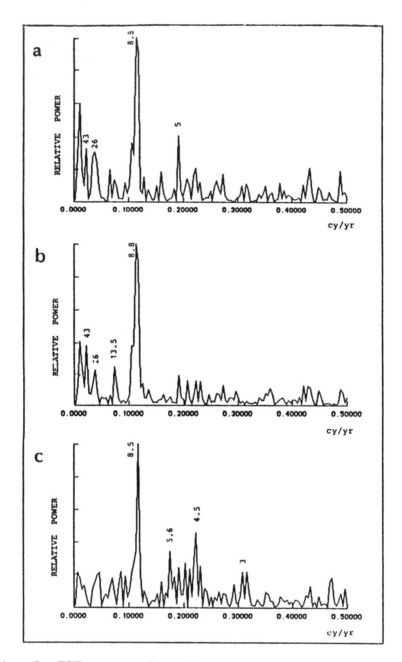

Fig. 5. FFT spectral analyses on the data from Lago
Blanco. (a) thickness of the varves, (b) thickness
of the summer units and (c) thickness of the winter
units.

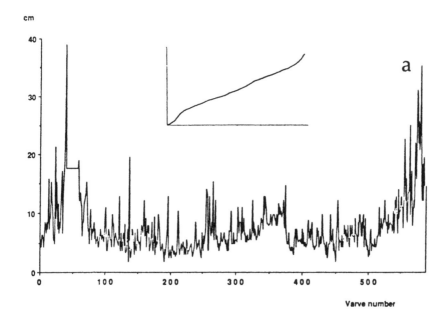

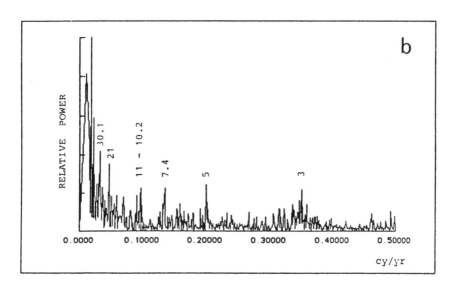

Fig. 6. Río Corintos. (a) Thickness of the varves, the insert represents the cumulative curve of the growth of the varves. (b) FFT spectral analysis on the thickness of the varves.

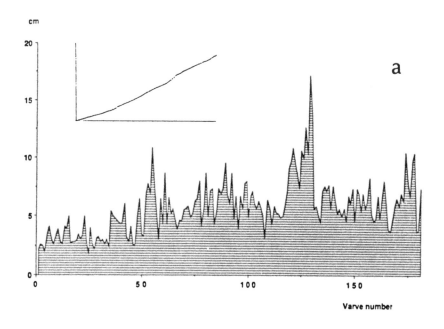

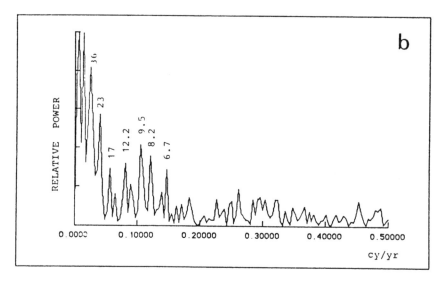

Fig. 7. Lago Epuyén. (a) Thickness of the varves, the insert represents the cumulative curve of the growth of the varves. (b) FFT spectral analysis on the thickness of the varves.

LAGO EPUYEN

The glacial history of this region has been very well documented by Caldenius (1932a), who measured in the depression of Arroyo del Carbón three main profiles (only the two, labelled 4 and 7, were published in detail). He interpreted the variation in the varve sequence as representing an ice recession calculated to be about 30-33 m/year, followed by a new readvance of the ice mass, responsible for the erosion of the uppermost part of the original sequence. Caldenius (1932a) assigned to this deposits to his "Gotiglacial" due to the fact that they are lying inside the third moraine-system.

Out of Caldenius' two published profiles, we chose the younger, according to his stratigraphy, i.e. number 7 (Figure 7a), because there are a number of missing varves (60) in profile 4.

FFT spectral power analysis of the series gives main concentration of density at 36, 23, 17, 12.2, 9.5, 8.2 and 7 years.

Because the 9.5- and 8.2-years cycles are of fairly importance here, it is difficult not to think about the similarity between these peaks and those obtained in Lago Blanco. At least both deposits are associated with a readvance of the ice, and their cumulative curves do not look unsimilar.

RIO VIZCACHAS

The moraines of Río Vizcachas-Tapi Aike region lie on distal and proximal glaciolacustrine deposits including very deformed silty and clayey varves. In a well-exposed section in the Rio Vizcachas Valley, Caldenius (1932a) found a 3 m thick sequence containing 101 probably Daniglacial varves, inbetween to till beds. The pattern of the variation of the thickness of the varves is shown in Figure 8a.

Although the sequence only covers 101 years, the short terms trends are well-defined. FFT spectral analysis (Figure 8b) gives major density for 18.3-, 12.8-, 6.7-, 5.5-, 3.4- and 2.2-years cycles.

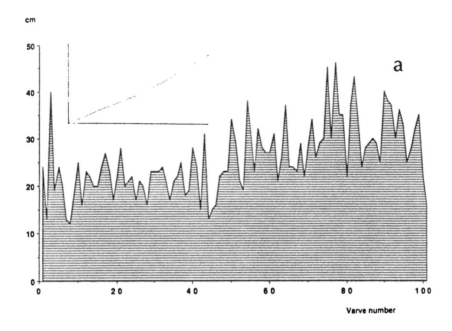

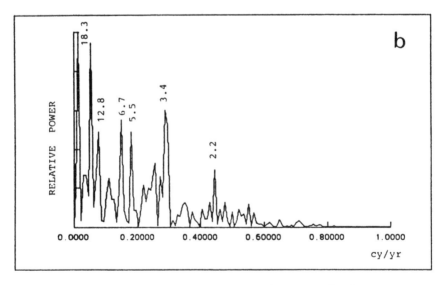

Fig. 8. Río Vizcachas. (a) Thickness of the varves, the insert represents the cumulative curve of the growth of the varves. (b) FFT spectral analysis on the thickness of the varves.

LAGUNA BLANCA

In Laguna Blanca, at the southernmost end of South America, Caldenius carried out detailed geochronological investigations. In the southeastern margin of the lake, he measured five main profiles, all of them containing very well developed sequences of glacial varves. His profile number 14, containing 383 distal varves (Figure 9a), was interpreted by varve chronology as being simultaneous with the varves from Lago Epuyén (Caldenius, 1932a). For this reason and, in addition, because it has not a single missing varve, we chose this profile to apply time series analysis.

FFT spectral power analysis (Figure 9b) shows relevant peaks corresponding to 26.9, 21.3, a concentration between 13.1 and 11.6 years, and minor terms at 8.7, 7.6, 6.1,4 and 3 years.

CONCLUSIONS

The results can be analyzed as derived from 1) data from winter and summer units, 2) data obtained from sequences deposited during glacial conditions and during deglaciation process, and 3) data from different localities.

1. The results obtained by using the time series analysis show that the individual measurements of summer and winter units give a more complete and useful information, as established by Agterberg and Banerjee (1969), and Schove (1979). The genetic difference between the summer units (deposited by turbidity currents) and the winter units (deposited out of suspension) is observed in the results from thickness-time series from Lago Buenos Aires and Lago Blanco.

2. Considering the sequence of Lago Buenos Aires as typical for the ice retreat, and the sequence of Lago Blanco as representing deposition during the last glacial maximum, two well-differentiated shapes were found:

a) During deglaciation the 3-years cyclicity is recorded by summer units, while, when full glacial conditions prevailed, the same term is recorded by winter units.

b) The 11- and 22-years cycles, attributed to the solar sunspot activity, are only recorded by the sequence representing deglaciation conditions.

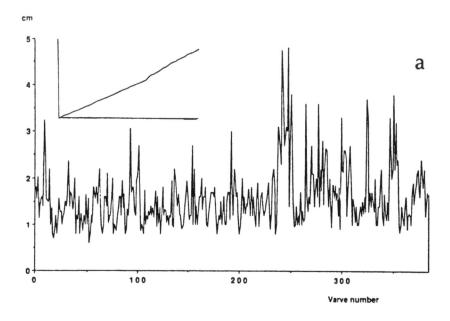

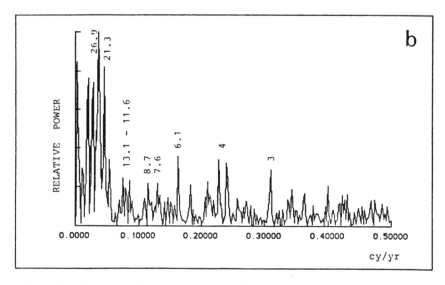

Fig. 9. Laguna Blanca. (a) Thickness of the varves, the insert represents the cumulative curve of the growth of the varves. (b) FFT spectral analysis on the thickness of the varves.

c) In Lago Buenos Aires, 18.2- and 11-years cycles were recorded. The 18.6-years lunar tidal cycle occurring in combination with the 11-years solar cycle has been recorded in several climatic indics during the Holocene, not only in South America but also worldwide (Fairbridge and Sanders, 1987; Currie, 1983, 1987). This is recorded only in the sequence representing deglaciation conditions. Furthermore, this similarity suggests that it is possible that these terms began to be recorded as early as about 13,000 years B.P., when the ice receded because of a climatic amelioration. If so, it may indicate that, at that time, climatic variations in Patagonia during the deglaciation phase were similar to the Holocene ones.

d) The strongly dense 8.8-/8.5-years cycle recorded in Lago Blanco (full glacial conditions) is not recorded in the data from deglaciation time. It must represent a strong climatic signal. There is no evidence about the nature of this climatic signal, however. Varve thickness is found to be sensitive to precipitation, temperature (Perkins and Sims, 1983) and river run-off (Granar, 1956).

These differences give a firm base for the study of older varve series, when data from individual summer and winter units become available. In this way, climatic information of the Pleistocene could be obtained and interpreted in a new way.

3. Regarding the time series analysis from the six different localities (analysis on the total varve thickness) we conclude that:

a) In all the records, unless the one from Lago Blanco, there is a conspicuous concentration of density either at around 11-years or at around 22-years or both, which is interpreted as reflecting the sun cyclicity.

b) The records from Lago Buenos Aires and Río Corintos show some significant similarities supporting the general correlation of the ice retreat in both localities.

c) On the other hand, the time correlation between Laguna Blanca and Lago Epuyén, argued by Caldenius (1932a), is not supported by time series analysis. However, there is a great distance between both basins, and the incidence of the west winds is different because Lago Epuyén is protected by the relatively high Andes.

d) The 3-years cycle (Schove, 1979) is present in four localities, being absent in Lago Blanco and Río Vizcachas.

Many questions, i.e. 8.5-/8.8-years cycle, the 2-years delay and the short term trends from 5- to 8-years cycles, remain unsolved. We trust, however, that with the help of future data these uncertainties will be understood.

ACKNOWLEDGEMENTS

The author would like to thank Prof. Fairbridge for encouraging conversations.

Dr Morner measured the sequences from Lago Buenos Aires and Lago Blanco, teaching the author (not only in Patagonia, but also in Sweden and Finland) the chronological method here employed. For this, and for many inspiring discussions, the author's gratitude.

REFERENCES

Agterberg, F.P. & Banerjee, I. 1969. Stochastic model for the deposition of varves in glacial Lake Barlow-Ojibway, Ontario, Canada. Can. J. Earth Sci., 6:625-652.

Anderson, R.Y. 1961. Solar-terrestrial climatic patterns in varved sediments. **New York Academy of Sciences Ann.**, 95:424-439.

Anderson, R.Y. & Dean, W.E. 1988. Lacustrine varve formation through time. **Palaeogeography, Palaeoclimatology, Palaeoecology**, 62:215-235.

Antevs, E. 1922. The recession of the last ice sheet in New England. **Am. Geogr. Soc.**, 2, 120pp.

Arrhenius, G. 1947. En studie over Uppsalatraktens varviga margel. **Sveriges Geologiska Undersokning**, Stockholm, 486(5):1-74.

Ashley, G.M. 1975. Rhythmic sedimentation in glacial Lake Hitchcock, Massachusetts - Connecticut. In: **Glaciofluvial and glaciolacustrine sedimentation**. Jopling & McDonald (eds.), Soc. Econom. Paleont. Mineral, special publication, 23:304-320.

Caldenius, C.C. 1932a. Las glaciaciones cuaternarias en la Patagonia y Tierra del Fuego. **Geografiska Annaler**, Stockholm, 1-2:1-164. Also published as: Dirección General de Minas, 95, 150pp., 1932, Buenos Aires.

Caldenius, C.C. 1932b. Las glaciaciones cuaternarias en la Patagonia y Tierra del

Fuego y sus relaciones con las glaciaciones del hemisferio boreal, estudio geocronológico. **Anales de la Sociedad Cientifica Argentina,** 113:49pp.

Currie, R.G. 1983. Detection of 18.6 year nodal induced in the Patagonian Andes. **Geophysical Research Letters,** 10:1089-1092.

Currie, R.G. 1987. Examples and implications of 18.6- and 11-yr terms in world weather records. In: **Climate, history, periodicity and predictability.** Rampino, Sanders, Newman & Koningsson (eds.), 378-403. New York, Van Norstrand Reinhold.

De Geer, G. 1912. A geochronology of the last 12,000 years. **Compte Rendu IX Cong. Geol. Intern.,** 1:241-258. Stockholm, 1910.

De Geer, G. 1940. Geochronologia Suecica Principles. **Kungl. Sven. Vetenskapakad. Handl.,** 18(6):367pp. Stockholm.

Fairbridge, R.W. 1984. Planetary periodicities and terrestrial climate stress. In: **Climatic changes on a yearly to millennial basis.** Morner & Karlén (eds.), D. Reidel Publishing Co., Dordretch. 509-520. Holland.

Fairbridge, R.W. & Sanders, J.E. 1987. The sun's orbit, A.D. 750- 2050: basis for a new perspective on planetary dynamics and Earth-Moon linkage. In: **Climate, history, periodicity and predictability.** Rampino, Sanders, Newman & Koningsson (eds.), 446-471. New York, Van Norstrand Reinhold.

Flint, R.F. 1971. **Glacial and Quaternary Geology.** 892pp. New York, John Wiley & Sons.

Granar, L. 1956. Dating of recent fluvial sediments from the estuary of the Angerman river (the period 1850-1950 A.D.). **Geol. Foren. Forhandl.,** 78(H4):654-658.

Hogbom, A.G. 1889. Om relationen mellan kalcium-och magnesiumkarbonat i de kvartara avlagringarna. Geol. Foren. **Forhandl.,** Bd II. Stockholm.

Ignatius, H. 1958. On the rate of sedimentation in the Baltic sea. **Extrait des Compte de la Société Geologique de Finlande,** 30:135-144.

Johnston, W.A. 1922. Sedimentation in Lake Louise, Alberta, Canada. **Am. J. Sci.,** Ser. 5, 23:376-386.

Keranen, R. 1984. Certain relationships between lake level variations and some climatic factors in Finland. In: **Climatic changes on a yearly** to millennial basis. Morner & Karlén (eds.), D. Reidel Publishing Co., Dordretch, 381-389. Holland.

Kuenen, P.H. 1951. Mechanics of varve formation and the action of turbidity currents. Geol. Foren. Forhandl., H1:69-84. Stockholm.

Menke, W. 1986. Scientist's Helper. User's manual. Macintosh version 2.11 for the 512K Macintosh Plus and Macintosh XL, 24pp.

Morner, N.A. 1978. Varves and varved clays. In: **The encyclopedia of Sedimentology**. Fairbridge & Burgeois (eds.), 841-843.Pennsylvania:Dowden, Hutchinson & Ross.

Morner, N.A. & Sylwan, C.A. 1987. Revised terminal moraine at Lago Buenos Aires, Patagonia, Argentina.IPPCCE Newsletters, 4:15-16.Japan.

Morner, N.A. & Sylwan, C.A. 1988. Magnetostratigraphy of the Patagonian moraine sequence at Lago Buenos Aires. **Journal of South American Earth Science** (submitted).

Nilsson, E. 1932. Quaternary glaciations and pluvial lakes in British East Africa. **Geografiska Annaler**, Stockholm, vol. 13.

Norin, E. 1927. Late glacial varves in Himalaya connected with the Swedish time scale. **Geografiska Annaler**, 9, 2. Stockholm.

Perkins, J.A. & Sims, J.D. 1983. Correlation of Alaskan varve thickness with climatic parameters, and use in paleoclimatic reconstruction. **Quaternary research**, 20:308-321.

Schove, D.J. 1979. Varve-chronologies and their teleconnections, 14,000-750 B.C. In: **Moraines and varves**. Schluchter (ed.), 319-325. Rotterdam:A.A. Balkema.

Schove, D.J. 1983. Introduction. In: **Sunspot Cycles**. Schove (ed.), 1-28. Pennsylvania:Hutchinson Ross.

Sonnet, C.P. & Williams, G.E. 1985. Solar periodicities expressed in varves from glacial Skilak Lake, southern Alaska. **Journal of Geophysical Research**, 90:12,019-12,026.

Stenborg, T. 1970. Delay of run-off from a glacier basin. **Geografiska Annaler**, 53:1-30.

Sylwan, C.A. 1987. Annual paleomagnetic record from late glacial varves in Lago Buenos Aires, Patagonia, Argentina. **Quaternary of South America and Antarctic Peninsula**, 5:181-196.

Sylwan, C.A. 1988. Paleomagnetism of glacial varves from the last glaciation maximum in Patagonia (Lago Blanco), Argentina.

Physics of the Earth and Planetary Interiors (submitted).

Terasmae, J. 1963. Notes on palynological studies of varved sediments. Journ. Sed. Petrol., 33:314-319.

Williams, G.E. & Sonnet, C.P. 1985. Solar signature in sedimentary cycles from the late Precambrian Elatina Formation, Australia. Nature, 318:523-527.

M.J.ORGEIRA, L.A.BERAZA, H.VIZAN & J.F.A.VILAS
University of Buenos Aires and CONICET, Argentina

M.L.BOBBIO
University of Buenos Aires, Argentina

8

Evidence for a geomagnetic field excursion in the Late Pleistocene (Entre Ríos, Argentina)

In memory of Prof. Daniel A. Valencio

ABSTRACT

Paleomagnetic and susceptibility data from a 2.5 m thick loessic and estuarine sedimentary sequence exposed near Gualeguaychú (Entre Ríos, Argentina) are reported. According to previous 14C data, estuarine conditions existed during the Late Pleistocene between 35,400 +/- 1,800 y.B.P. and 26,600 +/- 720 y.B.P. If the radiocarbon dates represent in fact minimum ages, these estuarine environments would have taken place during the last Interglacial, around 100 ka B.P. (Sangamon, Isotope Stage 5).

Natural remanent magnetization (NRM) directions obtained from 73 samples show that some of these have recorded a magnetic component different from the normal present geomagnetic field (Brunhes Epoch). The stability of the NRM was analyzed by alternating fields demagnetization (AF) and thermal demagnetization.

The stable remanence of the samples defined by thermal demagnetization suggests that some sections - mostly the estuarine ones - have recorded oblique polarities. AF demagnetization data show analogous results.

These results suggest that a geomagnetic field excursion was recorded in the sequence. On the basis of geologic and paleomagnetic data, the correlation between this excursion and the Laschamp Event (last geomagnetic event) as well as between this and other excursion of the geomagnetic field are discussed.

173

RESUMEN

Datos paleomagnéticos y de susceptibilidad magnética de una secuencia de sedimentos loésicos y estuáricos, de 2,50 m de espesor, expuesta en las cercanías de Gualeguaychú (Entre Ríos, Argentina), se presentan en este trabajo. De acuerdo a datos radiocarbónicos existentes condiciones estuáricas habrían existido en esta localidad durante el Pleistoceno Tardío entre 35.400 +/- 1.800 a A.P. y 26.600 +/- 720 a A.P. Si estos fechados representan en realidad sólo edades mínimas, estos ambientes estuáricos habríanse desarrollado durante el último Interglacial (Sangamon) alrededor de 100 Ka A.P. (Estadío Isotópico 5).

Las direcciones de magnetización natural remanente (MNR) obtenidas en 73 muestras demuestran que algunas de ellas han registrado un componente magnético diferente del campo magnético normal presente (Epoca Brunhes). La estabilidad de la MNR fue analizada por demagnetización de campos alternantes (AF) y por demagnetización térmica.

La remanencia estable de las muestras definidas por demagnetización térmica sugiere que algunas secciones -principalmente las estuáricas- han registrado polaridad oblicua. Los datos de demagnetización por AF muestran resultados análogos.

Estos resultados sugieren que una excursión del campo geomagnético fue registrada en la secuencia. Sobre la base de datos geológicos y paleomagnéticos, se discute la correlación entre esta excursión y el Evento Laschamp (último evento geomagnético) así como entre ésta y otras excursiones del campo geomagnético.

INTRODUCTION

In the past few years geoscientists all over the world have been specially interested in the discovery of polarity events and geomagnetic field (GF) excursions during the Late Cenozoic. On one hand, studies carried out in different parts of the world are aimed at analyzing the behaviour of the GF; on the other hand, it is particulary important to determine the extent - whether global or not - of these events and excursions, since they could be used as a tool to establish geological correlations at the regional or the global levels.

In 1982, a paleomagnetic sampling was carried out in an outcrop near the harbour of Mar del Plata, Argentina, approximately 2 m (7 ft) thick, of loessic sediments, calcareous material and poorly-consolidated sands and containing marine fossils (Orgeira, 1983). The outcrop´s geological location suggests that these sediments correspond to a transgression in the Late Pleistocene. The results of the paleomagnetic studies performed on samples taken from this profile led to the suggestion that in some of the analyzed samples a remanent magnetic component was present showing a polarity practically opposite to that of the present geomagnetic field (Orgeira, 1983; 1985). For this reason, a decision was made to find other sedimentary sequences of equivalent geological age in order to confirm the existence, in South America, of a record of a geomagnetic event or a geomagnetic field excursion during the mentioned period.

On the basis of the stratigraphical and geochronological data obtained from the study area, Guida and Gonzalez (1984) suggested the existence of a paleoestuary in the southeast of the Entre Ríos province, between 35,400 +/- 1,800 and 26,000 +/- 720 y.B.P. These radiocarbon dates could be interpreted as minimum ages as well, thus suggesting instead a Last Interglacial (Sangamon, Isotopic Stage 5) age for the bearing sediments. In view of this information this area was chosen for the study.

PALEOMAGNETIC STUDIES

A paleomagnetic sampling was carried out in a sedimentary sequence assigned to the Late Pleistocene near the city of Gualeguaychú (Entre Rios) (Figure 1). The studied profile (Figure 2), approximately 2.5 m (8 ft) thick and the top of 0.9 m (3ft) corresponding to the present soil layer, consists of loessic deposits and estuarine sediments containing invertebrate marine fossils.

Seventy - three oriented samples were collected in order to study the stability of the magnetic remanence in the sedimentary sequence; out of the total number, 33 of them were put under a demagnetization treatment by the exposure to high temperature. Vertical equidistance between successive samples varies between 0 and 5 cm (2

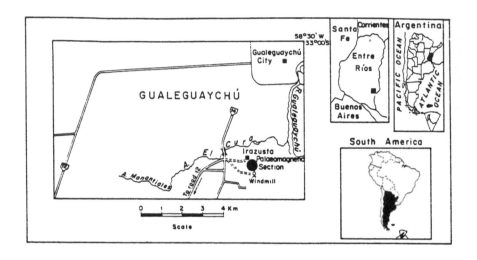

Figure 1. Location of geomagnetic sampling.

inches). Similarly, magnetic cleaning by
alternating fields (AF) was applied to the
remaining 40 samples to verify the results obtained
for the thermal cleaning. Another forty specimens,
equidistant throughout the sequence, were collected
separately to measure the sequence´s magnetic
susceptibility.

As a first step, sample magnetic
susceptibility was measured; Figure 2 shows the
values obtained in relation to the stratigraphic
position of the samples.

Then, laboratory tests were carried out to
measure natural remanent magnetism (NRM) in all
collected samples; Figure 3a shows the directions
of the samples prepared for the thermal cleaning.
Note that there are several specimens in which NRM
directions are remarkably different from the
direction of the present geocentric axial dipolar
magnetic field. This suggests that the NRM in these
samples would include at least one component having
a direction different from that of the geomagnetic
field prevailing during the last 0.73 my (Bruhnes
Epoch). On this basis, the traditional techniques
of thermal and magnetic cleaning were applied in
order to isolate stable remanent magnetizations
(SRM) in the specimens and/or identify possible
overprinted magnetizations in them.

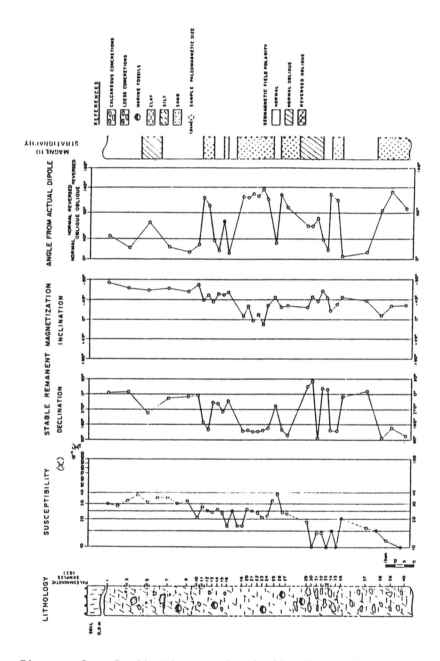

Figure 2. Declination and inclination of the stable remanent magnetization (SRM) and angle between the VGP for each stratigraphic level and the actual geomagnetic dipole. The magnetostratigraphy, the lithology and the susceptibility (X) logs for this sequence are also shown.

177

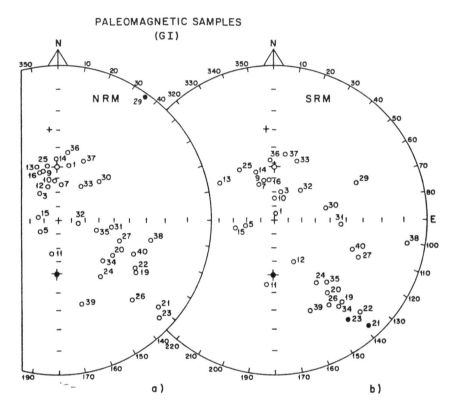

Figure 3. Direction of natural (3a) and stable (3b) remanent magnetization for the samples cleaned by thermal demagnetization. The direction of the actual geomagnetic dipole (crossed open circle) and its antipodal (crossed solid circle) are shown. Negative inclination (open circle); positive inclination (solid circle).

Temperatures used for the thermal cleaning were 100, 150, 200, 250, 300, 350, 375, 400, 450, 475 and in some cases - where it was possible - 500oC; after each cleaning stage the residual remanent magnetism (RRM) of the specimens was measured.

Figure 4a shows direction variations in the RRM and a relative intensity decrease after each cleaning stage in specimen GI 21; the SRM was isolated after the 375oC for this specimen.

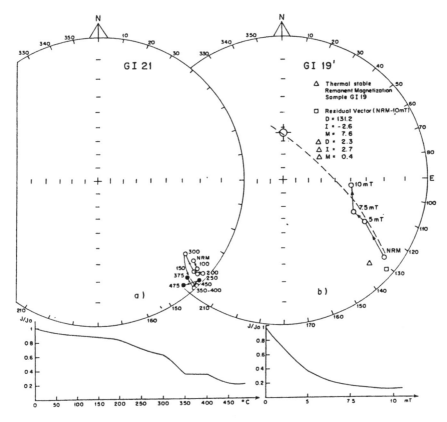

Figure 4. a) Changes in direction and intensity (J/Jo) of magnetization by thermal cleaning for specimen GI21; b) changes in direction and intensity (J/Jo) of magnetization by progressive AF cleaning for the specimen GI19′.

D=declination   1=inclination   M=intensity

Captions as in Figure 3.

Note that directions corresponding to the isolated SRM still differ from the present geomagnetic field direction. The magnetic behaviour of this specimen with relation to thermal cleaning is representative, sensu lato, of that observed in the other specimens, which showed NRM directions remarkably different from the present magnetic field direction.

Figure 3b shows the SRM isolated from all the samples exposed to thermal demagnetization. Again, as it can be easily seen here, there are still several specimens with SRM directions different from the direction of the present geocentric axial dipolar magnetic field; similarly, it should be pointed out that almost all the samples having this peculiar magnetic remanence belong to paleoestuarine sediments (Figure 2 and 3b).

The magnetic cleaning procedure consisted in applying peak magnetic fields of 5, 7.5, 10, 12.5 and 15 n1, RRM being measured after each step. The magnetic behaviour of specimens thus treated was consistent in some cases with the results obtained for the thermal cleaning. In other cases, where the NRM direction departed from that of the present dipolar geomagnetic field, magnetic behaviour was different; it was observed that the directions of the RRM isolated in the successive cleaning stages define, for each specimen, a maximum circle containing approximately the direction of the present dipolar geomagnetic field and that of the corresponding NRM. The existence of this maximum circle suggests that two GF directions have been recorded - an oblique one: remanent magnetism 1, RM 1; and another one coinciding with the present dipolar geomagnetic field: remanent magnetism 2, RM 2 - NRM being the resultant of both. The isolated SRM tends to coincide with the present dipolar geomagnetic field direction (RM 2); this seems to imply that, through the successive cleaning stages, the other component (RM 1) has been gradually eliminated. With the purpose of isolating this latter component, vector subtraction was applied to each NRM and RRM obtained after the 10 nT cleaning stage (optimal cleaning). In each case the direction resulting from this vector subtraction coincides, sensu lato, with the thermally isolated SRM for the equivalent specimen. This type of behaviour is illustrated in Figure 4b (RRM intensity and direction variations for specimen GI 19′, after each magnetic cleaning stage); this figure also shows the direction of the vector subtraction result (NRM - RRM 10 nT) and the SRM direction isolated through thermal cleaning in the equivalent stratigraphic sample (GI 19).

It follows that, in these samples, the AF cleaning technique mainly cleans the oblique direction component (RM 1) recorded in the specimen thus treated. Therefore, thermal cleaning proved to be the most suitable method for isolating this component (RM 1) in the samples.

INTERPRETATION OF THE RESULTS

Figure 3b suggests that the illustrated SRM directions are arranged with preference into two main groupings. In order to define these groups more clearly, the isodensity lines of the SRM directions obtained through thermal cleaning were drawn on an equal-area net (Figure 5). This drawing confirms, in fact, the existence of two well-defined groupings (A1 and A2). One of them (A2) shows a maximum density core practically coincident with the direction of the present dipolar geomagnetic field. Note that it coincides with RM 2, defined in the previous section; the other one (A1) has its maximum density core approximately in the following coordinates: D 140o, I -20o (approximate direction of component RM 1, defined in the previous section).

There are intermediate directions between these two maximum density groupings. According to the thermal cleaning paths observed in some of these specimens (see example in Figure 6a), these intermediate directions might indicate, in each case, a direction resulting from the overlapping, to various degrees, of two magnetic components having directions corresponding to the previously mentioned groupings (a1: RM 1; A2: RM 2). Moreover, other samples, that have SRM directions which fall between A1 and A2, show no evidence of the existence of more than one magnetic component (Figure 6b). Consequently, these intermediate directions could also be interpreted as a record of the GF for a given moment in the sequence.

The SRM directions of each of the main groupings (A1 and A2) were used to calculate the average values corresponding to each population (Figure 7). Note that A1 directions are grouped in an elliptic arrangement; that might be due to the fact that these directions consist of one magnetization (RM 1), plus small and varying proportions of a second magnetization approximately coincident with that of the present dipolar geomagnetic field (RM 2). Similarly, this peculiar arrangement of directions could also be attributed to the behaviour of the GF during the period under study.

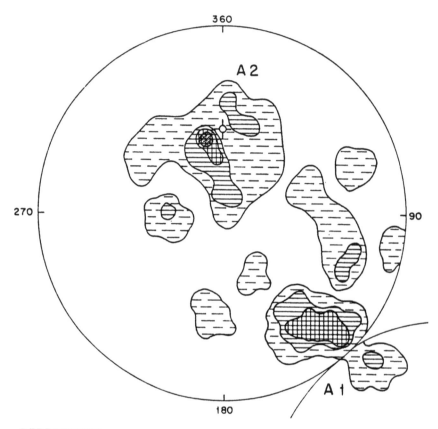

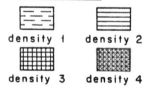

density 1   density 2

density 3   density 4

Figure 5. Equal area net for the direction of stable remanent magnetization (SRM) obtained by thermal cleaning. Captions as in Figure 3.

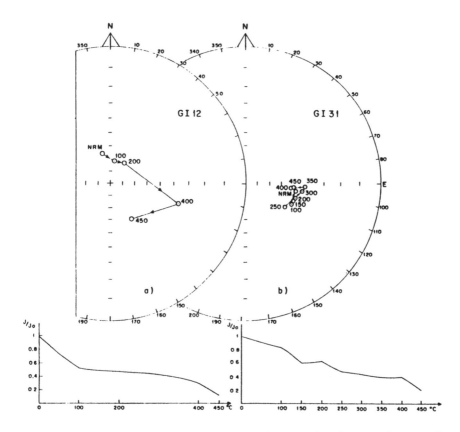

Figure 6. Changes in direction and intensity of magnetization by thermal cleaning of specimen GI12 (6a) and GI31 (6b). The specimen GI12 shows evidence of two magnetic components and in the specimen GI31 only one magnetic component is present. Captions as in Figure 3.

The SRM directions obtained through thermal cleaning were used to calculate the virtual geomagnetic poles (VGP) for each of the stratigraphic levels studied (Table 1). Each VGP´s angle from the present dipole was also determined (Valencio et al., 1977). This is illustrated in Figure 2 together with the corresponding magnetic stratigraphy. Observing the values obtained for the magnetic stratigraphy it can be seen that the sedimentary sequence has recorded two positions of different polarity for the GF. One of them, having

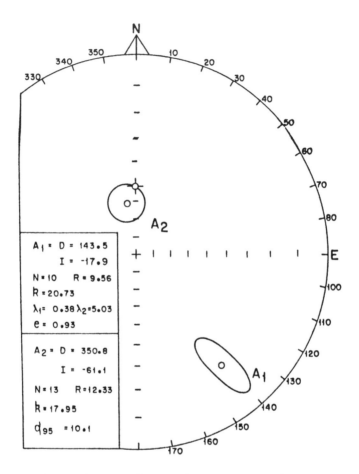

Figure 7. Mean direction of stable remanent magnetization (SRM) of the specimens located in A1 and A2.

D=declination I=inclination N=number of samples

R=length of the resultant of n unit vectors

K=Fisher precision parameter

1 and 2 = minimum and maximum axis of the ellipse of 95% confidence, respectively.

95 = radius of the circle of 95% confidence about the mean direction.

e=ellipticity

normal-normal oblique polarity, has been recorded in the upper section of the sequence. The other one, having reverse oblique polarity, is more commonly found recorded in the middle (paleoestuarine sediments) and lower sections of the sequence.

In accordance with the succession of polarity records in the sediments, a sequence of magnetization acquisition can be suggested for the studied specimens which showed evidence of the recording of two magnetic components (RM 1, RM 2). The primary magnetization in these specimens would be RM 1, with reversed oblique polarity, recorded during the sedimentary process. RM 2, with normal polarity, seems to have been overprinted later, possibly when the upper part of the sequence was deposited.

CONCLUSIONS

a) The Curie temperatures determined in the thermally treated specimens suggest that the minerals responsible for the magnetic remanence in the sequence would belong to the titano-magnetites series.
b) The observed SRM of reversed oblique polarity (RM 1) are recorded, mainly, in the estuarine sediments containing marine fossils ranging in age from 35,400 +/- 1,800 to 26,600 +/- 720 y.B.P. perhaps Mid-Wisconsin in age (Guida and González, 1984). However, these radiocarbon dates could be interpreted as minimum ages, suggesting instead a last Interglacial (Sangamon, Isotope Stage 5) age for these marine sediments, which would be consistent with our present knowledge of paleoeustatic curves. In any case, this magnetization would represent an GF excursion during the Late Pleistocene, being reported for the first time in South America.

For an age equivalent to the above-mentioned Mid-Wisconsin alternative, an GF excursion has been reported in Lake Mungo, southeastern Australia (Barbetti, M. and Mc Elhinny, M., 1972). Figure 8 shows the VGP resulting from the archaeomagnetic study carried out in this Australian locality, and the average VGP calculated for grouping A1 (RM 1, with reversed oblique polarity) in this study (Gualeguaychú VGP, G1), together with the corresponding confidence ovals. A comparison of

Table 1

Sample	Position of virtual north pole	
GI1	42oN	120oE
GI3	67oN	103oE
GI5	18oN	163oE
GI7	68oN	176oE
GI9	74oN	171oE
GI10	61oN	122oE
GI11	30oS	130oE
GI12	18oS	102oE
GI13	39oN	204oE
GI14	71oN	200oE
GI15	15oN	172oE
GI16	78oN	159oE
GI19	32oS	64oE
GI20	30oS	81oE
GI21	40oS	60oE
GI22	33oS	67oE
GI23	47oS	62oE
GI24	27oS	87oE
GI25	58oN	210oE
GI26	35oS	82oE
GI27	11oS	58oE
GI29	28oN	33oE
GI30	26oN	54oE
GI31	10oN	55oE
GI32	53oN	61oE
GI33	71oN	19oE
GI34	35oS	77oE
GI35	25oS	79oE
GI36	84oN	252oE
GI37	79oN	351oE
GI38	8oS	40oE
GI39	40oS	96oE
GI40	7oS	57oE

these VGP leads to the conclusion that the Lake Mungo VGP -30,780 y.B.P. (Figure 7)- might be correlated with the Gualeguaychú VGP (G1) and both might correspond to the same geomagnetic excursion.

Taking into consideration the geographical distance separating Gualeguaychú from Lake Mungo, and the proximity existing between G1 and F7, the GF would show an important dipolar component contribution during the reported excursion. This excursion, in turn, does not appear to be related to the Laschamp Geomagnetic Event (Bonhommet, N. and Zahringer, J., 1969), also assigned to the Late Pleistocene, because the polar positions are not coincident (Figure 8).

Figure 8. VGP position of Laschamp (L), Olby (O) (France); Lake Mungo (S1, S6, S7, S8, F6, F7, F8) (Australia); Gualeguaychú (G1, G2) (Argentine). G1 and F7 have been represented with their 95% confidence ellipse.
Data of Laschamp and Olby from Bonhomet and Zahringer (1969).
Data of Lake Mungo from Barbetti and Mc Elhinny (1972).
open circle = in the shown hemisphere.
solid circle = in the opposite.

ACKNOWLEDGEMENTS

The authors wish to thank the Universidad de Buenos Aires, and the Consejo Nacional de Investigaciones Científicas y Técnicas (CONICET) for the support which enabled this work. They also thank the useful suggestions from the scientific members of Laboratorio de Paleomagnetismo de la Universidad de Buenos Aires. Special thanks to Lic. E. Oviedo.

REFERENCES

Barbetti, M. & Mc Elhinny, M. 1972. Evidence of a geomagnetic Excursion 30,000 y.B.P. **Nature,** vol. 239, (5371):327-330.

Bonhommet, N. & Zahringer, J. 1969. Paleomagnetism and potassium argon age determinations of the Laschamp Geomagnetic Polarity Event. **Earth and Planetary Science Letters.** 6. 43-46. Amsterdam.

Guida, N.G. & González, M.A. 1984. Evidencias paleoestuáricas en el sudeste de Entre Ríos, su evolución con niveles marinos relativamente elevados del Pleistoceno superior y Holoceno. **IX Cong. Geol. Arg., Actas III,** 577-594., S.C. de Bariloche.

Orgeira, M.J. 1983. **Informe final de la Beca de estudio.** Comisión de Investigaciones Científicas de la Pcia. de Buenos Aires, La Plata, unpublished report.

Orgeira, M.J. 1985. **Informe final de la Beca de Perfeccionamiento.** Comisión de Investigaciones Científicas de la Pcia. de Buenos Aires, La Plata, unpublished report.

Valencio, D.A.; Vilas, J.F. & Mendía, J.E. 1977. Paleomagnetism of sequences of red beds of the middle and upper sections of the Paganzo Group (Argentine) and the correlation of upper Palaeozoic-Lower Mesozoic rocks. **Geophys. J. Roy. Astr. Soc.,** 51(1):50-74.

M.T.PROST
ORSTOM, Cayenne, French Guiana

9

Beaches and cheniers in French Guiana

ABSTRACT

The French Guiana Holocene coastal plain is an open
ocean chenier plain. The shoreline is composed of
extensive shoreface attached mudflats and
characterized by waterfront mangroves and by sub-
coastal swamps and marshes. There are presently six
migrating mudbanks along the French Guiana's coast,
each 20 to 40 km long. When a mudbank is attached
to the coast, the shoreline is undergoing
progradation. Within the interbank zones, on the
contrary, shoreline is undergoing erosion and sandy
deposits create specific types of accretionary and
erosional coasts. In the inner part of the coastal
plain, narrow sandy ridges are observed thanks to
aerial photos. They are isolated, perched and
shallow-based sandbodies resting on marine clays,
i.e. typical cheniers.

This paper deals with the coastal area between
Cayenne and the Maroni River. Our aim is to provide
information concerning sandy material of beaches
and cheniers and to discuss some points related
with them. After a brief review of the dynamic and
the morphology of the shoreline, the significance
of chenier's evolution is presented within the
Guianese's environment. The characteristics of the
sandy formations are shown within three key-areas.
General results are discussed in conclusion.

RESUMEN

La planicie costera holocena de la Guayana Francesa
es una planicie de chenier, océnica abierta. La
línea de costa está compuesta por planicies
fangosas extensivas yuxtapuestas a la costa y
caracterizadas por manglares, pantanos y marismas
subcosteras. Hoy, al presente, seis planicies
fangosas migratorias a lo largo de la costa de la
Guayana Francesa, cada una de ellas de 20 a 40 km
de largo. Cuando una planicie fangosa se yuxtapone
a la costa, la línea de la ribera está sufriendo
progradación. Dentro de las zonas interbancos, por
el contrario, la línea de costa está registrando
erosión y los depósitos arenosos crean tipos
específicos de costas acrecionales y erosivas. En
la porción interior de la planicie costera, crestas
arenosas angostas son observadas gracias a fotos
aéreas. Ellas son cuerpos arenosos aislados,
colgados, poco profundos, que descansan sobre
arcillas marinas, por ejemplo, cheniers típicos.
Este trabajo se ocupa del área costera entre
Cayena y el rio Maroni. Nuestro objetivo es proveer
información concerniente a materiales arenosos de
playas y cheniers y discutir algunos puntos
relacionados con ellos. Luego de una breve revisión
de la dinámica y la morfología de la línea de
costa, se presenta la significancia de la evolución
de los cheniers dentro del ambiente guyanés. Las
características de las formaciones arenosas se
muestran en tres áreas clave. En las conclusiones
se discuten los resultados generales.

INTRODUCTION

The French Guiana Holocene coastal plain, called
the "young coastal plain", faces the open
equatorial Atlantic Ocean. It has an altitude which
ranges between 0 and 5 m and stretches along 320
km. Located between the Oyapock River (Brazil
border) and the Maroni River (Suriname border), it
is part of the 1600 km long open ocean chenier
plain of the Guianas region which itself stretches

from the Amazon mouth in Brazil to the Orinoco delta in Venezuela. This coastal area presents similar dynamic conditions and morphological features.

This paper deals with the coastal area between Cayenne and the Maroni River where cheniers have been identified thanks to both aerial photos interpretation and field observations. The I.G.C.P./201 Project (International Geological Correlation Program: Quaternary of South America. IUGS/UNESCO) has promoted this research since 1984. Currently, research has been taken over by the ORSTOM-Center of Cayenne, under the Coastal Environment Programme.

MUDBANKS AND SANDY RIDGES

Present-day shoreline changes in French Guiana are directly linked to the huge sediment discharge of the Amazon River that supplies 250 million m3/y transported along the Guiana's coast. A "blanket" of mud covers the nearby continental shelf at least down to the isobath of 20 m and, on average, the shelf is 30 km away of the shoreline (Bouysse et al., 1977; Jeantet, 1982; Pujos and Odin, 1986). As a result, this coastal system presents an inverse sedimentological transition (nearshore muds to offshore sands) of the "typical coastal system" defined by Reineck and Singh (1986). Offshore sands in French Guiana nearby continental shelf are relict deposits.

One part of the Amazon discharge (110 millions m3/y) is carried steadily westward along the nearshore by the combined action of the equatorial current and the longshore currents created by the trade-wind-waves (Wells and Coleman, 1977; Rine and Ginsburg, 1985). Consequently, vaste migrating mudbanks are formed, specific of the Guiana's coastline. In French Guiana there are presently six migrating mudbanks each 20 to 40 km long (Froidefond et al, 1985). When a mudbank is attached to the coast, the coastline is undergoing progradation. Within the interbank zones, on the contrary, shoreline is undergoing erosion. These phenomenons are alternative and simultaneous. They occur in very short-term periods and shoreline changes are really striking (Prost, 1986). Stacks of sediments from migrating mudbanks create on the shoreface a vertical sequence of laminated and

massive muds and, on the coastal plain, a dynamic
horizontal sequence of mud marshes and sand
cheniers (Rine and Ginsburg, 1985).

Muddy environment appears as dominant not only
along the shoreline but also on the nearby
continental shelf. In comparison, sandy material
appears as rather rare. However, Augustinus (1978)
has clearly shown how important are sand deposits
concerning the geomorphological evolution of
shoreline. They create specific types of
accretionary and erosional coasts within the
interbank zones. Moreover, the narrow sandbodies
observed in the inner part of the coastal plain
thanks to aerial photos, are recorded with former
coastlines. The study of sandy environments, in
conclusion, is significant for the present-day
shoreline changes and for the understanding of the
Holocene coastal plain evolution.

THE CHENIERS

Two major systems of sandy ridges have been
recognized in French Guiana: "old" ridges are
situated between the marine clays of the Mara Phase
(8000/6000 BP) and those of the Moleson Phase
(2600/1300 BP); "recent" ridges are separating
Moleson sediments from those of the Comowine Phase
(1000 BP to present). "Old" and "recent" ridges are
disposed roughly parallel to the present shoreline.
The geomorphological disposition of the ridges
is an insufficient field indicator (Seurin, 1975);
their degree of pedogenesis, though more accurate,
cannot be the only data. As regard to this problem
we are able to present the following given points:
-Holocene ridges are isolated, perched and shallow-
based sandbodies resting on the Demerara clays,
i.e. typical cheniers.
-Cheniers disposition must be rather complicated,
particularly within the estuaries environment as in
the key-areas of Mana and Sinnamary (Prost, 1986;
Lointier, 1986).
-Present-day shoreline cheniers formation continues
to occur within the interbank area, in a narrow
zone around high water level (key-area: Cayenne).
Ridges migrate westwardly due to beach-drifting.
-When more sand is removed than it is deposited
(key-area: Pointe Isere), cheniers are eroded,
particularly during spring tides and during the
high wind period (January to May).

Two major problems are linked with chenier formation:
1) The first one concerns the origin of the sand. The original theory was that chenier sand was supplied by local rivers. Krook (1968) and Augustinus (1978) reported that the chenier sand in the eastern part of the young coastal plain of Suriname originates from the Maroni and from the coast of French Guiana, owing to mineralogical data. However, they also emphasize that sometimes this sand has a local admixture of sand from the pelite deposits off the coast. As regard to this problem, Krook defined two heavy mineral associations within Suriname sandy coastal area: a) the staurolite bearing sand is supplied by the Maroni and the French Guiana coast; b) the epidote-hornblende sand probably originates from the Amazon River.
2) The second point involves the problem of the shoreline evolution. What are the respective parts of sea-level changes and progradation and retreating sequences?. A rapid sea-level rise between 10000 BP and 6000 BP is a given point (Brinkman and Pons, 1968; Turenne, 1978). But we know very little about the post-Mara evolution. On one hand, cheniers are considered as built up on a relatively stable coast (Augustinus, 1978). On the other hand, no part of the Earth crust can be considered as stable and the relationship between isostatic movements and eustatic changes is very complex. This problem is specially significant in Suriname where the coastal plain is part of the Berbice subsidence area.

This paper only deals with the first point. The aim is to provide more detailed information about the sandy morphological units and material of French Guiana.

CHARACTERISTICS OF THE SANDY FORMATIONS

It is known that sediment grain parameters - as grain-size distribution, mineral composition, shape, roundness and grain surface texture - are environmental indicators. Such studies present, of course, several limitations and need more data to be completed; nevertheless, they can provide interesting information. Grain-size distribution is controlled by the hydro-dynamic conditions existing exactly at the moment of deposition; shape, roundness and grain surface texture provide a more long-term data.

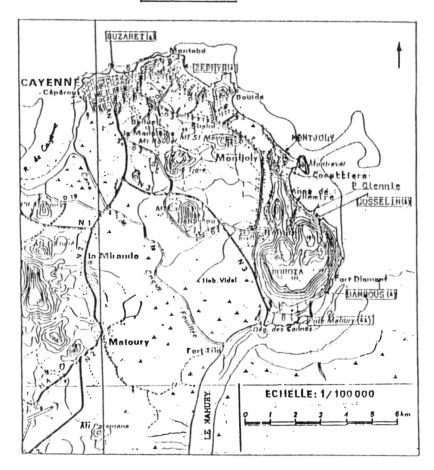

▲- Areas where samples were collected

Figure 1. Sketch map of Cayenne area.

To obtain information concerning present and Holocene sandy material, samples were first collected along the beaches of Cayenne and then compared with other samples taken from beaches and cheniers situated between Sinnamary and the Maroni River. Laboratory's analyses were carried out at the Center-ORSTOM of Cayenne and supervised by J.L. Duprey; they concerned grain-size distribution, including histogram and cumulative curves and the values of Qdphi, So and He. Shape, roundness and grains surface texture were analyzed with a binocular microscope. Approximately 900 grains were counted and a great number have been drawn. The results of the counts have been represented graphically. Some graphs have been place on one sheet for correlation purposes.

1 CAYENNE KEY-AREA

The "Isle of Cayenne" is one of the rare points of the coast where the rocks of the shield are in contact with the ocean. As a result, the coastline is a succession of rocky points and isolated hills separing sand coves and straight beaches (Figure 1). Between the Mahury River mouth, in the east, and the Cayenne River mouth, in the west, there is a sequence of sandy formations, with good accessibility. The samples were collected the same day and under the same tidal conditions, mostly in the foreshore area and a few in the topstrata of the backshore ridges. This east-west sequence is also a dynamic one: beaches of the Rorota coast are situated presently within an interbank zone; beaches near Cayenne town center, on the contrary, are fixed behind the mudbank and the mangrove and, as a result, out of the wave action.

1.1 GRAIN-SIZE DISTRIBUTION

Table 1 presents the mean grain-size and the compared values of Qdphi, So and He values. It defines sedimentation units taken in 1987 (May 31th). Sands are mostly coarse-grained and composed of almost pure quartz. Shell clastics are rare and appeared only near some rocky points (Buzaret). Black concentration of heavy minerals is found within some foreshore area and linked with the backwash of the waves (pointe Glennie; Gosselin beach).

Table 1. Median particle diameter and Qdphi, He and
So values.

Beaches of CAYENNE

Beaches and ridges	Md	Qdphi	He	So
ZEPHYR	0,195 mm	O,2	O,19	1,16
COCOTTIERS	0,315 mm	0,15	0,10	1,12
GOSSELIN	0,355 mm	0,2	0,2	1,12
MAHURY (ridge)	0,360 mm	0,15	0,12	1,12
MONTJOLY (ridge)	0,340 mm	0,3	0,3	1,24
MONTJOLY (foreshore)	0,222 mm	0,375	0,35	1,30
ANSE DE REMIRE	0,320 mm	0,375	0,2	1,30
BUZARET	0,365 mm	0,45	0,4	1,38
BAMBOUS	0,290 mm	0,5	0,3	1,42
MAHURY (ridge)	0,250 mm	0,625	0,5	1,55

Three kinds of sorting have been distinguished:
-Sorting is excellent (=1,1/1,2) at the foreshore
area of Zephyr, Cocottiers and Gosselin, and at the
crest of the backshore ridges of Montjoly and
Mahury. Cumulative curves have a straight form (as
an S) which indicates a homogeneous material
(Figure 2).
-Sorting is good (=1,3/1,4) at the foreshore of
Montjoly and Anse de Remire, slightly less
homogeneous than the previous samples. Cumulative
curves are also less straigth (Figure 2).
-Sorting is not so good (more than 1,4) at the
foreshore area of Buzaret and Bambous and mostly at
the newly formed beach of Mahury point. Cumulative
curves revealed a more heterogeneous material
(Figure 2).

Figure n° 2

KEY-AREA CAYENNE: Cumulative curves

Grain size distribution

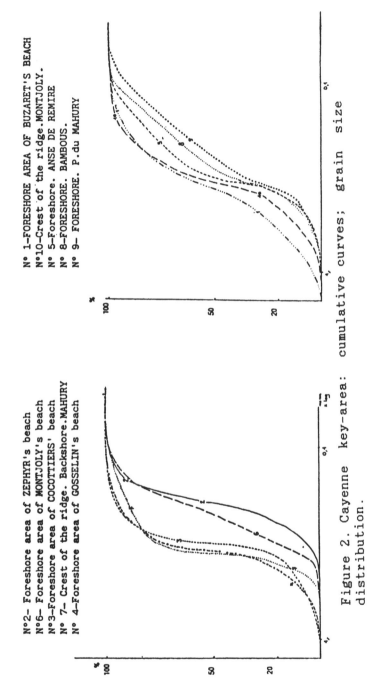

N°2- Foreshore area of ZEPHYR's beach
N°6- Foreshore area of MONTJOLY's beach
N°3-Foreshore area of COCOTTIERS' beach
N° 7- Crest of the ridge. Backshore.MAHURY
N° 4-Foreshore area of GOSSELIN's beach

N° 1-FORESHORE AREA OF BUZARET'S BEACH
N°10-Crest of the ridge.MONTJOLY.
N° 5-Foreshore. ANSE DE REMIRE
N° 8-FORESHORE. BAMBOUS.
N° 9- FORESHORE. P.du MAHURY

Figure 2. Cayenne key-area: cumulative curves; grain size
distribution.

Table 2. Surface texture of sand grains.

BEACHES OF CAYENNE

BEACHES	ROUNDNESS %	WELL ROUNDED %
Buzaret	58,4	6,7
Zéphyr	62	1
Montjoly (estran)	79	16
Montjoly (cordon)	81,4	15,4
Cocottiers	79,8	24,8
Anse de Rémire	77,2	22,8
Gosselin	80,8	16,8
Bambou	71	21,5
P. du Mahury (estran)	70	6,4
P. du Mahury (cordon)	73	22

BEACHES AND SANDY RIDGES (W of CAYENNE)

BEACHES	ROUNDNESS %	WELL ROUNDED %
Isère 1	85	5,5
Isère 2	78,8	15,5
Isère 3	90,4	25
Isère 4	83,7	9,8
Van Uden 1	80,3	22,7
Van Uden 2	82,7	26
Amarante	83,5	7,4
Goulet	89,6	28
Piste Hattes	81,2	11
Route 8	85,8	16,1
Trou Poisson	82,5	15,1
Sinnamary	71,7	2,2

In conclusion, grain-size distribution shows that:
-Foreshore sands and backshore ridges are formed by the same material, though the topstrata of the ridges has a better sorting. The latter is related with wind action, as we observed in the field.
-Sands have a good sorting. Qdphi, He and So values are respectively less than 0,6; 0,5 and 1,5.
-Sorting of sandy material that is accumulating within the interbank zone is roughly better from the east to the Cayenne River mouth.

These analyses, as well as our field observations, allow us to consider the Mahury River as the major source of sandy material that is accumulating along the Cayenne's shoreline.

1.2 SHAPE AND SURFACE TEXTURE OF GRAINS

Six classes for roundness determination were used: angular (NU = non-usés); sub-angular (Sub-Ang. = Sub-angulaire); sub-rounded (CA = coins-arrondis); rounded (Arr = arrondi); ovoid (Ov = Ovoide) and well-rounded (R = rond). Sands of Cayenne beaches have good roundness, 80 to 52% of grains rounding (Table 2). But when we compared this percentage with those of the last three classes (Arr, Ov and R, i.e. grains with the best rounded = "bien faconnés"), it appeared that they do not exceed, in the best situation, 25% of the sample. In some cases, as in Zephyr, they represent only 1% of the sandy material. This means that roundness is still in an early stage. Most of the grains are sub-angular and sub-rounded.

Sand grains of the backshore's ridges show a better rounding as compared with beach sands, feature linked with a selective action of the wind. The roundness of sand grains also shows a relationship with grain size: coarse fractions (1 mm; 0,800 mm; 0,630 mm) have in general better roundness than finer ones.

Concerning the surface texture of sand grains (Figure 3), we distinguished five kinds of grains: angular and sub-angular bright grains (EN = éclat naturel); smooth shining grains (EL = émoussés luisants); smooth grains with shining and frosted surface (PL, picotés-luisants), smooth frosted grains (M = mats) and opaque grains (OP).

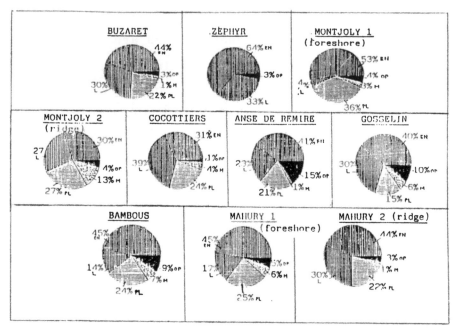

FIGURE N° 3

SURFACE TEXTURE OF SAND GRAINS.

BEACHES OF CAYENNE. May 1987.

—EN – Angular and subangular shining grains (éclat naturel)
—L – Smooth shining grains (émoussés-luisants)
—PL – Smooth grains with shining and frosted surfaces.
     (picoté-luisant)
—M – Rounded frosted grains (mats)
—OP– Opaque grains

Bright grains are common (from 31 to 64% of samples), indicating neighbouring of the supply and its immaturity. On the opposite, the EL grains form between 15 to almost 40% of the sample; a reworking, with the mixture of sand grains is very likely. The most significant part of quartz grains is composed by PL grains. Frosted grains (M) are little, linked with the topstrata of backshore ridges.

Opaque grains show heavy mineral concentrations and polymineral grains. They are linked with the basement rocks. The hills of Cayenne belong to the Ile de Cayenne Formation (quartzites, granites, migmatites, etc) and to the Paramaca Formation (volcanic rocks) (Figure 4). In the hinterland the shield is composed by the schistes of the Orapu and the Bonidoro Formations. All these formations supplied heavy minerals. Staurolite and disthene, for instance, appeared in the alluvial deposits of little "creeks" and valleys in the north of the Montagne des Chevaux, and tourmaline in those of Montsinery-Tonegrande area.

The last point - but not the least - is that sandy formations of Cayenne shoreline are mostly coloured. The ochre colour of the sands is produced by a sort of frosted or water polished patine. Quartz grains present a ragged and tapped surface, with deep holes (as tooth caries) filled with dark red-brown coating. Some grains seem corroded. These characteristics are those of a weathering formation (weathering shield profiles and pedogenesis).

In short:
- Shoreline sands show an average roundness, with a majority of sub-angular and sub-rounded grains. A reworking of residual sediments is, however, very likely, owing to the percentage of smooth shining grains.
- The roundness of grains shows a relationship with grain-size and with the selective action of the wind.
- Sands are composed by almost pure quartz with a lower percentage of opaque grains and heavy minerals. A shield supply and a fluvial transport are significant concerning the heavy mineral composition.
- Coloured and tapped quartz grains indicate a weathering action.

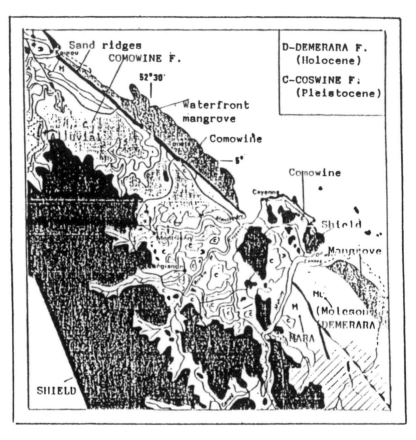

1/500.000

(modified of GUILLOBEZ S., 1979)

Figure 4. Geological sketch.Key-area Cayenne-Kourou

Sand formations that are accumulating along
the Cayenne coast are very likely supplied by the
Maroni River and they are reworking in a marine
environment.

2 Mana-Pointe Isere Key-area

We compared the results of Cayenne with the sands
of some beaches and cheniers of the Mara-Pointe
Isere key-area (Figure 5).
    This area has been studied in detail since
1984 in the framework of the IGCP-201 Project and
by the Hydrology-Geomorphology Department of the
ORSTOM-Center of Cayenne (Lointier and Prost, 1987;
Prost, 1986; 1987).

2.1 SAMPLES OF THE GALIBI BEACH. POINTE ISERE

The Galibi beach is situated on the seaward side of
the Pointe Isere, in the site of an old Galibi
village now deserted. In 1955, the Pointe Isere was
very large owing to a prograding muddy coast.
Presently, it is the site of a strong erosion.
    Samples have been taken under the same
physical conditions in January 1987 during spring
tide. They concern the foreshore surface, the
receding chenier of the backshore area, the
toplamina of the latter and the washover sand of
the landward side of the chenier (Figure 6).
    Four cumulative curves concern the Galibi
area. Samples 87-6 and 87-7 have similar curves.
Sample 87-4 is more homogeneous and sample 87-5 is
the finest one (Figure 7b).
    Sample 87-6 was taken on the surface of the
washover fan - very coarse-grained material, eroded
from the coastal sand and distributed inland by the
waves of spring tide, during high winds and heavy
sea period. Such kind of material indicates a
typical straight erosion coast. Sample 87-7
concerns the topstrata of the receding chenier. It
presents a significant heavy mineral concentration
owing to deflation (mostly sherry coloured grains).
The roundness of these samples is good (more than
80%) but it is obvious that washover sands have a
more heterogeneous stock concerning the surface
texture of grains (Figure 8).

KEY AREAS: POINTE ISÈRE, MANA, ORGANABO-SINNAMARY (modified of GUILLOBEZ, 1979)

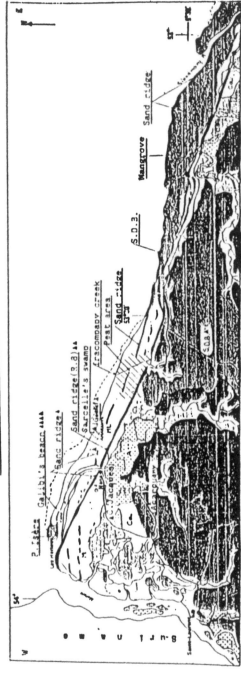

1/500.000

DEMERARA FORMATION (Holocene) - Waterfront mangroves (Comowine phase)
ML - Swamps and marshes (Wleson phase)
M - Fresh water swamps (Mara phase)

COSWINE FORMATION (Pleistocene) - Cs - Sandy Coswine (offshore bar landscape)
Ca - Clayed Coswine (Clay's landscape)

A - Alluvial deposits   S.D.3-Série Détritique de Base

▲ - Areas where samples were collected

Figure 5. Sketch map.

204

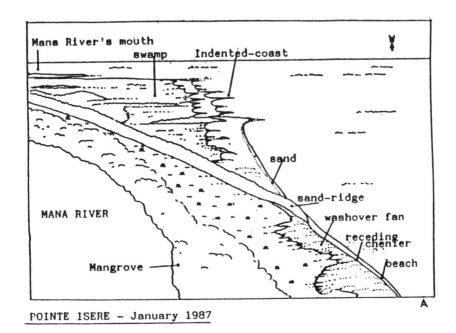

POINTE ISERE - January 1987

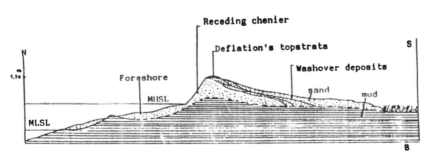

Pointe Isere. French Guiana.

Figure 6. Partial view of Pointe Isere and Schematic cross-section of Galibi Beach.

The other two curves (87-4 and 87-5) concern sands of the receding chenier and of the foreshore area. The latter is the finest. The percentage of the best rounded grains is more significant than within previous samples (sample 87-4, 15,5%; sample 87-5, 25%).

To obtain further information we compared these samples with those of some neighbouring cheniers (Figure 7b). Samples 87-3 ("recent" sandy ridge within Van Uden's rice fields) and 87-8 (road Mana-Aouara, km 1) show cumulative curves superposed to washover sands curve. Sand is also very coarse-grained. The roundness is good (more than 80%) and the percentage of the best rounded grains sometimes significant – 23% within the sample 87-3, i.e. almost the same value as in the foreshore of Galibi beach.

2.2 SAMPLES FROM NEIGHBOURING BEACHES: AMARANTE AND GOULET

Some samples have been taken more eastbound along the Sarcelle swamp shoreline. The swamp area is 20 km long and 4 km wide (Figure 5). The free-water central area has 450 ha, with an average depth of 25 cm (July 1985). Sarcelle swamp is separated from the sea by a straight coast with very narrow beaches and by the waterfront mangrove. One part of the shoreline was under mud progradation in 1984-1985 (Goulet and Amarante beaches enclosed). Currently, mud is going out and these beaches are undergoing erosion.

Cumulative curves show homogeneous material (Figure 9a) with an S form. Comparison with the material of the receding chenier of Galibi point shows similar features; sands have the same granulometrical parameters. Roundness and grains surface texture are also comparable (Figure 8). The percentage of opaque grains is relatively high, mostly at Amarante foreshore (19%); sherry-coloured grains form 16% of the sample, very likely staurolite. It is interesting to notice that sherry-coloured grains have frosted surface, mostly at the coarser fractions (1 mm; 0,800 mm; 0,630 mm). However, these dull features do not cover all the grain, and bright parts (looking like shock marks) often appeared.

206

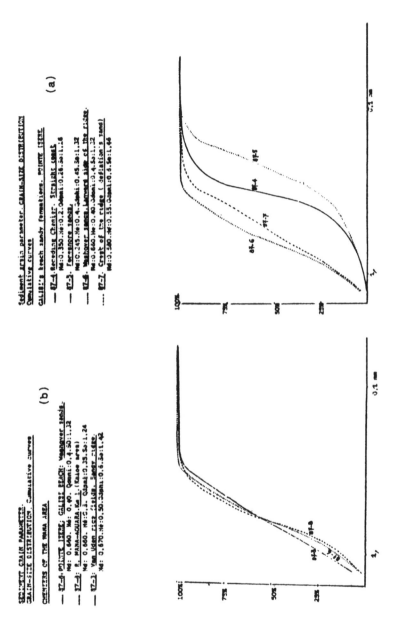

Figure 7. Key-area: Mana. Cumulative curves

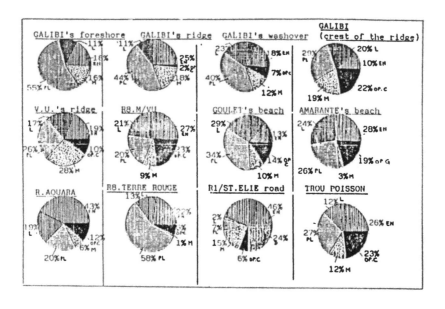

- ▥ – <u>EN</u>. Angular and subangular shining grains (éclat naturel)

- ▦ – <u>L</u>. Smooth shining grains (émoussés-luisants)

- ▦ – <u>PL</u>. Smooth grains with shining and frosted surfaces.

- ▨ – <u>M</u>: Rounded frosted grains (mats)

- ▦ – <u>D</u>. Grains with an etched surface (dépolis)

- ■ – <u>OP.C.G.</u> – Opaque, sherry couloured and pink grains
  (opaques, caramels translucides et grenats
  translucides)

Figure 8. Surface texture of sand grains. Key-
areas: Pointe Isere, Mana, Sinnamary

These results were compared with sandy
formations of some cheniers situated a few
kilometres away from the shoreline, behind the
Sarcelle swamp (Figure 9b). Only one example is
presented here; it concerns the sands situated 2 m
deep over silty clays, along the junction channel
between the Mana River and Van Uden rice fields
(road no 8). Sands are well sorted, pure quartz,
with high concentration of sherry-coloured and pink
grains (very likely staurolite and garnet), a
typical Mana-Maroni heavy mineral association
(Krook, 1968; Augustinus, 1978).

3 SINNAMARY KEY-AREA

To conclude, these results were finally compared
with sandy material of two "old" ridges along the
no 1. The first is a sand quarry situated at the
crossing of road no 1 and Saint Elie road (Figure
10). Sand is very white. Cumulative curve is
similar to those of several sandy ridges at the old
coastal plain (Mana region). The 24% of the grains
have an etched surface, characteristics that are
similar to those of the S.D.B. sand quarry (km 149,
road no 1).
    The ridge of Trou Poisson, on the contrary,
shows similar features to those of the inner
coastal cheniers (Sarcelle's swamp area) (Figure
9b).

GENERAL RESULTS

1. Comparison of the mean grain-size of all samples
shows that sandy material is, in general, coarse-
grained (Figure 11). Very few samples concern
medium sands. Fine sand and shell clastic cheniers,
known in Suriname (Krook, 1968; Augustinus, 1978),
have not been found here.
2. Sands present a good sorting. Comparison of the
Qdphi, So and He values (Figure 12) presents
roughly the same features.
3. Relationship between rounded sand grains and
well-rounded sand grains (Figure 13) shows that
roundness is not too developed, with a majority of
sub-angular and sub-rounded grains. That seems
obvious owing to the migrating mudbanks and the
low-to-medium energy of the marine environment.

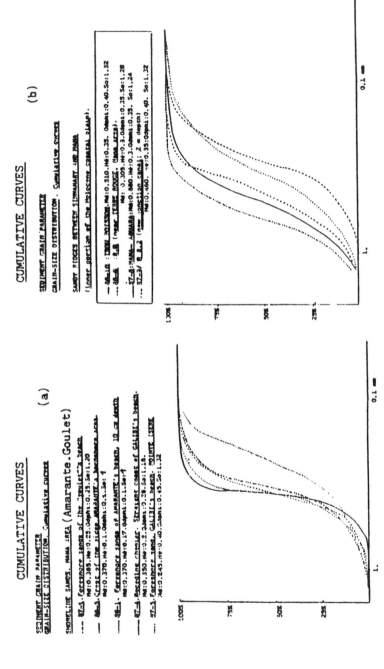

Figure 9. Sandy formations (W of Cayenne). French
Guiana coast.

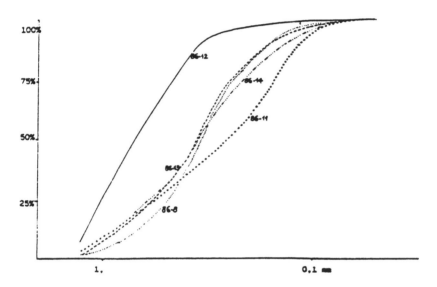

GRAIN-SIZE DISTRIBUTION. Cumulative curves

Sandy ridges of the inner part of the young coastal
plain, or behind it.

— 86-12: CRIQUE JACQUES (Mana).1.30 m depth). Ochre sands.
        Md:0,7.Hé:0,5.Qdphi:0,55.So:1,42
--- 86-13/R.ACAROUANY. 30 m alt. Grey sands
        Md:0,370.Hé:0,5.Qdphi:0,625.So:1,42
..... 86-11. R.l. km 149. White sands
        Md: 0,569. Hé:0,75.Qdphi:0,95.So:1,68
___ 86-14. R. Mana-St. Laurent, 11 m alt. Brown sands
        Md:0,350.Hé:0,7.Qdphi:0,9.So:1,36
.... 86-8. R l. Sinnamary area (Near R St. Elie).White sands
        Md: 0,350.Hé:0,5.Qdphi:0,5.So:1,42

Figure 10. Sediment grain parameter.

4. Reworking of residual formations (older cheniers
for instance) is very likely if we considered the
part of the smooth shining grains, typical marine.
5. Heavy minerals concentrations appeared along the
coastline. In the west coast they form dark red-
brown deposits (Mana, Les Hattes area). Krook
(1968) and Augustinus (1978) report that almandite
and staurolite bearing sands of the east coast of
Suriname are supplied by the Maroni and the
neighbouring coast of French Guiana. A

concentration of sherry-coloured and pink grains (very likely staurolite/garnet association), has been found in the Mana zone.

6. Weathering is observed owing to the coloured and tapped surfaces of quartz grains of Cayenne and thanks to the etched surfaces of quartz and some heavy minerals (Sinnamary and Mana key-areas). Krook (1968) also reports that decreasing of garnet within chenier sands indicates the increase of weathering; the absence of garnet of the Wanica Phase in Suriname (Post-Mara and Pre-Moleson) is significant if compared with garnet percentage of the Moleson and Commowine ridges.

7. Figure 14 shows the relationship between Qdphi and the Md. We can observe that sandy material from Cayenne beaches and from shoreline formations of the west coast have similar dispositions. A relationship appeared between the washover deposits of Pointe Isere and the chenier of the Mana-Aouara road. Cheniers of Van Uden area are correlated with Pleistocene sands. At least, and "old" ridge of the Pleistocene coastal plain has similar position in the graph as the S.D.B. sands of the road no 1 (km 149). So, the supplies of S.D.B., Coswine sands and shield formations are evident.

CONCLUSION

These results show that:
- Chenier sands are (and were) supplied by local sources (basement weathering formations, S.D.B. Formation, Coswine Formation) transported by rivers and reworked by the sea. Reworking of old cheniers is also very likely.
- Until now, we have no data concerning sandy Amazon supply, as in Suriname. However, heavy minerals studies must be done to provide mineralogical data. Such information is necessary to discuss weathering problems and to define cheniers generations.
- Chenier evolution is very dynamic. It is (and always was) directly correlated with the specific shoreline conditions of the Guiana coast.
- Chenier formation continues to occur presently along the coastline. Accretionnary sandy coasts are very clearly expressed within the Cayenne Key-area (for instance, at Anse de Remire and at the eastern part of Montjoly beach). Erosional sandy coasts, on the contrary, appeared mostly on the west coast (Pointe Isere Key-area).

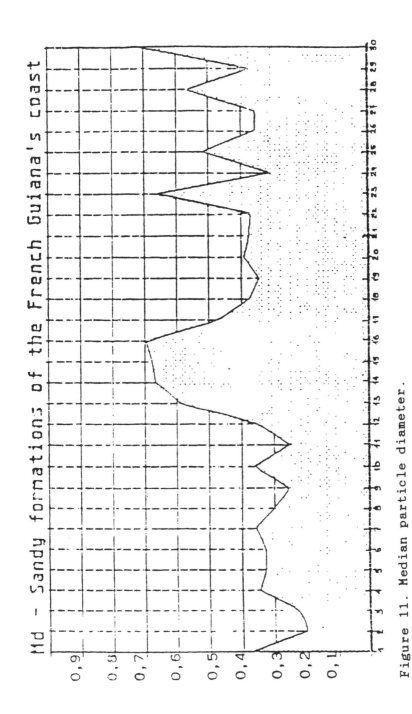

Figure 11. Median particle diameter.

213

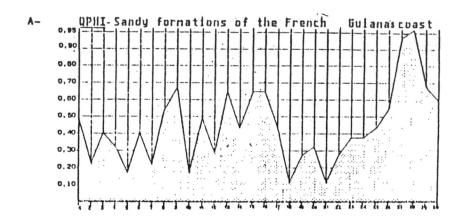

A— QPIII- Sandy formations of the French Guiana coast

B — QdPIII, SO and HE values (comparison)

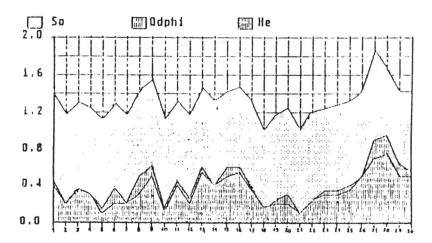

Figure 12. QPHI. Sandy formations of the French Giuana coast.

214

FORESHORE ZONE	CREST OF THE RIDGE	CHENIER
1. BUZARET's beach		
2. ZEPHYR's beach		
3. MONTJOLY's beach		
4. _____	4. MONTJOLY's ridge	
5. COCOTTIERS' beach		
6. ANSE DE REMIRE's beach		
7. GOSSELIN's beach		
8. DANROUS's beach		
9. MAHURY's beach		
10. _____	10. MAHURY's ridge	

KEY AREA: POINTE ISERE

11. GALIBI's beach		
12. Receding ridge. GALIBI's beach		
13. _____	11. GALIBI's ridge	
14. Washover sand. GALIBI's beach		

KEY-AREA: MANA/MARAIS SARCELLE

15. _____	15. _____	15. VAN UDEN's ridge
16. _____	16. _____	16. " " (10 cm depth)
17. _____	17. _____	17. New channel (2 m depth)
18. AMARANTE's beach (10 cm depth)		
19. AMARANTE's beach (20 cm depth)		
20. AMARANTE's beach (30 cm depth)		
21. _____	21. AMARANTE's ridge	
22. COULET's beach		
23. _____	23. _____	23. R. AOUARA (Koloc)
24. _____	24. _____	24. R B., TERRE ROUGE
27. _____	27. _____	27. R. MANA-ST. LAURENT 11 m alt.
29. _____	29. _____	29. R. ACAROUANY. 30 m alt.
30. _____		30. CRIQUE JACQUES

KEY-AREA SINNAMARY-ORGANABO

25. _____	25. _____	25. TROU POISSON
26. _____	26. _____	26. R.1 / ST. ELIE

28. SAND PIT OF THE S.D.B. FORMATION. (Serie Detritique de Base). Km 149, R. 1.

Key area: Cayenne. Key of figures 11 and 12

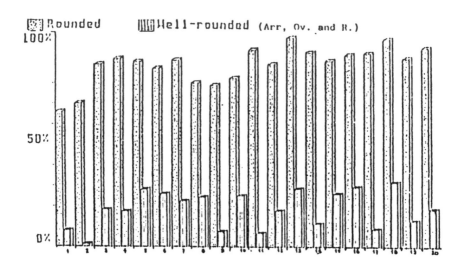

French Guiana's coast

| KEY-AREA: CAYENNE |

1. BUZARET's beach
2. ZEPHYR's beach
3. MONTJOLY's beach
5. MONTJOLY's: crest of the ridge
5. COCOTTIERS' beach
6. ANSE DE REMIRE's beach
7. GOSSELIN's beach
8. DANROUS' beach
9. MAHURY's beach
10. MAHURY: crest of the ridge.

| KEY-AREA: POINTE ISERE |

11. Foreshore zone of the GALIBI's beach
12. Receding chenier of the GALIBI's beach
13. Crest of the ridge (deflation's surface). GALIBI's beach
14. Washover sand. Landward side of the ridge. GALIBI's beach

| KEY-AREA: MANA-MARAIS SARCELLE |

15. Chenier. Van Uden's rice fields
16. New junction channel between the Mana River and VAN UDEN's rice fields. 2 m depth.
17. Foreshore zone of the AMARANTE's beach (10 cm depth)
18. Foreshore zone of the COULET's beach.
19. R. Mana-Aouara (Kaloe's rice fields)
20. R.G/TERRE ROUGE's road

Figure 13. Relationship between rounded sand grains and well-rounded sand grains.

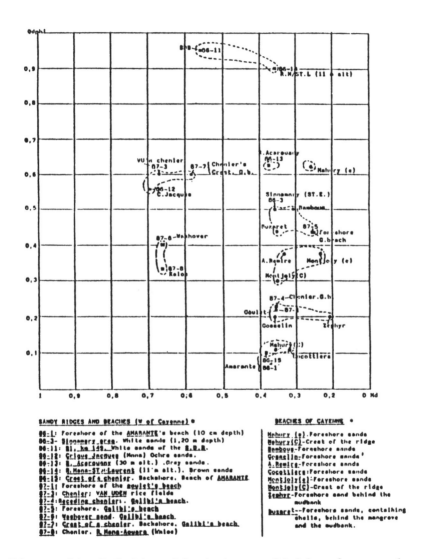

SANDY FORMATIONS. FRENCH GUIANA'S COASTLINE

Relationship between Qdphi values and the Md.

Figure 14. Relationship between Qdphi values and
the Md.

ACKNOWLEDGEMENTS

I gratefully acknowledge the laboratory of the Center ORSTOM (Cayenne) and in particular Mr. J.L. Duprey, that supervised the sediment analysis, and Mr Miatti, for their contribution in the preparation and study of grain-size distribution. I wish to thank Mrs A.C. Lhomme for the critical reading of the English text, and Mrs Bron that kindly typed the original original manuscript.

REFERENCES

Augustinus P. 1978. The changing shoreline of Surinam. Ultgaven "Natuurwetenschappelijke Studiekring voor Suriname en de Nederlandse Antillen. Utrecht, no 95. 232 p., 69 fig., 33 tabl., 17 pl., 234 ref.

Bouysse P.H., Kudrass H.R., Le Lann F. 1977. Reconnaissance sédimentaire du plateau continental de la Guyane Francaise (mission Guyamer, 1976). Bull. BRGM, Sect. IV, no 12. 141:179,20 fig. 4dpl.ht.

Brinkman R. & Pons C.J. 1968. A pedogeomorphological classification and map of the Holocene sediments in the coastal plain of the three Guianas. Soil Survey Inst. Wageningen, 25pp., 2 maps, 6 graphs.

Froidefond J.M., Prost M.T. & Griboulard R. 1985. Etude sur I´évolution morpho-sédimentaire des littoraux argileux sous climat equatorial: I´exemple du littoral guyanais. Rapport CORDET-IGBA. 189pp. Min. de la Recherche et de la Technologie. Univers. de Bordeaux-I. Depart. de Géol. et Océanographie. Talence.

Jeantet D. 1982. Processus sédimentaires et évolution du plateau guyanais au cours du Quaternaire Terminal. These 3eme cycle. Départ. Géol. Océanog. (I.G.B.A). Université de Bordeaux-I. Talence.

Krook L. 1968. Sediment petrographical studies in northern Suriname. Thesis Academish Proefscrift. Vrije Univ. Amsterdam.

Lointier M. 1986. Hydrodynamique et morphologie de I´estuarine de la Sinnamary. Le Littoral Guyanais: fragilité de I´environnement. Nature Guyanaise. SEPANRIT. Cayenne.

Lointier M. & Prost M.T. 1986. Morphology and hydrology of an equatorial swamp: example of the Sarcelle swamp in French Guiana. Quaternary of South America and Antarctic Peninsula. Vol. 4. K. Suguio, L. Martin and J. Rabassa. Rotterdam:A.A.Balkema.

Prost M.T. 1986. Morphologie et dynamique cotieres dans la région de Mana. Le littoral Guayanais: fragilité de l´environnement. Nature Guayanais. SEPANRIT. Cayenne.

Prost M.T. 1987. Aspects of the morpho-sedimentary evolution of French Guiana´s coastline. Quaternary of South America and Antarctic Peninsula. Vol. 4. K. Suguio, L. Martin and J. Rabassa. Rotterdam:A.A.Balkema.

Pujos M. & Odin J.C. 1986. Processus sédimentaires et évolution du plateau continental de la Guyane francaise au cours du Quaternaire Terminal. Univ. de Bordeaux-I. I.G.B.A. Projet CORDET A 22/A 73.

Reineck H.E. & Singh I.B. 1986. Depositional sedimentary environments. Springer-verlag. 551pp., 683 fig. New York.

Rine J.M. & Ginsburg R.N. 1985. Depositional facies of a mud shoreface in Suriname, South America. A mud analogue to sandy shallow-marine deposits. Journal of Sedimentary Petrology Vol. 55. 633:652.

Seurin M. 1975. Etude d´un cordon littoral a I´Anse de Remire (Guyane francaise). Travaux et Doc. de Géog. Trop. Talence. No 22. 207:216, 5 fig., 1 map, 1 photo h.t., 7 réf.

Turenne J.F. 1978. Sedimentologie des plaines cotieres. Atlas de la Guyane. CNRS/ORSTOM.

ANA MARIA BORROMEI
Universidad Nacional del Sur, Bahia Blanca, Argentina

10

A braided fluvial system in Pleistocenic sediments in southern Buenos Aires Province, Argentina

ABSTRACT

In Bajo San José (Buenos Aires Province), on the left bank of Sauce Grande River, excellent outcrops offered the possibility of studying detailed stratigraphic profiles of fluvial sediments belonging to the Late Pleistocene. According to Miall (1977), the following lithofacies were identified: Gm, massive or poorly bedded gravel; Gp, planar crossbedded gravel; Sp, planar crossbedded medium to very coarse sand; St, trough crossbedded fine to coarse sand; Sh, fine to very coarse sand with horizontal lamination in planar bed flow of upper or lower flow regime; Fl, very fine sand or silt, with fine lamination, ripples or massive. It is concluded that these deposits have been generated by a braided fluvial system.

RESUMEN

En la localidad de Bajo San José, Provincia de Buenos Aires, la presencia de excelentes afloramientos sobre la margen izquierda del Río Sauce Grande, permitió el estudio de perfiles estratigráficos detallados de sedimentos fluviales pertenecientes al Pleistoceno Tardío. Según lo establecido por Miall (1977), se reconocieron las siguientes litofacies: Gm, gravas macizas o mal estratificadas; Gp, gravas estratificadas con estratificación cruzada planar; Sp, arenas medias a muy gruesas con estratificación cruzada planar; St, arenas gruesas a finas con estratificación cruzada

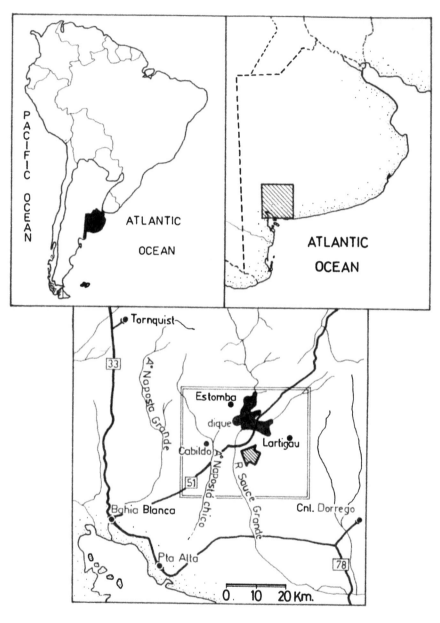

Figure 1. Location map of the studied locality, Bajo San José, southern Buenos Aires Province, Argentina

en artesa; Sh, arenas finas a muy gruesas con
laminación horizontal en flujo de lecho plano de
alto o bajo régimen de flujo; Fl, arenas muy finas
o limos con laminación fina, ondulitas o macizos.
Por lo tanto, se interpreta que estos depósitos han
sido generados a partir de un sistema fluvial
entrelazado.

INTRODUCTION

The purpose of this paper is to report new data
about the sedimentology of outcropping Quaternary
units located in the middle portion of the Sauce
Grande River Valley (Buenos Aires Province).
    A description of the sedimentary facies and
the fluvial system from which they were originated
is presented. The studied sequence is situated in
Bajo San José, along the provincial route N o
51 , 50 km northeast of the city of Bahía Blanca
and on the left bank of the Sauce Grande River
(Figure 1).
    Field work was performed during the geological
mapping of the "Cabildo" Sheet (IGM 1:100.000 N
o 3963-12), included in the Utilitary
Geological Economical Chart of Buenos Aires
Province.
    Likewise, this paper is also a contribution to
Project IGCP-201 "South American Quaternary".

STRATIGRAPHY AND AGE

The sediments of the studied unit are forming a
fluvial terrace -at 120 m a.s.l. (Figure 2) -,
composed of coarse, fine gravel and sand. These
sediments correlate with the Lower Psephitic Member
of the Agua Blanca Formation (Rabassa, 1985), of
Late Pleistocene age (Rabassa, 1985; Borromei,
1985).
    This unit is lying uncomformably on the
Pliocene -Lower Pleistocene substratum of the
Saldungaray and La Toma Formations (Furque, 1973).

METHODOLOGY

For the study of these Quaternary deposits, several
columnar stratigraphic profiles were surveyed at a
scale of 1:30 along different outcrops of the
quarry (Figure 3).

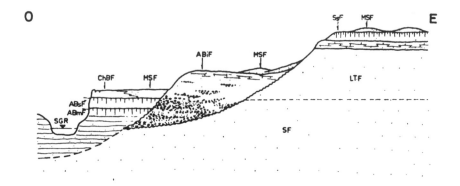

Figure 2. Transversal stratigraphic profile of the
Sauce Grande River Valley, in Bajo San Jose, Buenos
Aires Province. SF: Saldungaray Formation
(Pliocene), LTF: La Toma Formation (Early
Pleistocene), ABiF: Agua Blanca Formation Lower
Member (Late Pleistocene), ABmF: Agua Blanca
Formation Middle Member (Late Pleistocene), ABsF:
Agua Blanca Formation Upper Member (Holocene), SsF:
Saavedra Formation Upper Member (Holocene), ChBF:
Chacra La Blanqueada Formation (Holocene), MSF:
Matadero Saldungaray Formation (Holocene), SGR:
Sauce Grande River.

     The identification and description of the
existing facies as well as their associated
lithology and structures were based on the criteria
presented by Miall (1977).

DESCRIPTION OF THE FACIES

1 GRAVEL FACIES

Massive Gravel Facies (Gm)

This facies includes the massive or poorly
stratified conglomerates. The fabric is grain-
supported with poorly-sorted, subangular to
subrounded gravel, whose grain-size distribution
ranges from fine gravel to boulders. The prevailing
lithological type in the main clasts is quartzite
from the Sierras Australes range. The matrix is
sandy to sabulitic, poorly consolidated,

occasionally loose and with irregular cementation by chemically deposited iron and manganese. In some sections, imbricated clasts and pelitic intraclasts have been observed. These pelitic interclasts come from lower beds or the valley margins. There are intercalations with levels of medium to coarse, massive or horizontally laminated lenticular sand, which correspond to facies Sh. The presence of openwork gravels that alternate with sandy-sabulitic matrix gravels, represent cycles of deposition with high and low energy, where psammitic material infiltrated the crevasses of the gravel when flow energy decreased (Smith, 1974). These greyish brown beds are lenticular, over an erosive base and the observed thickness ranges from 0.30 to 1.70 m.

Planar cross-bedded Gravel Facies (Gp)

This facies is made up of grain-supported conglomerates with planar cross-bedding. With erosive base and top, these beds are lenticular, in which clast size varies from fine to coarse gravel and the matrix is coarse to medium sand with occasional cementation by chemically deposited iron and manganese. Thickness ranges between 0.30 to 0.70 m. The prevailing colour is light greenish grey.

2 SANDSTONE FACIES

Trough Cross-bedded Sandstone Facies (St)

This facies includes poorly consolidated and occasionally loose medium to very coarse sand, with trough cross-bedding. It is frequent the occurrance of lenses forming aligned or isolated quartzite pebbles. Beds are lenticular, dicordant, of a variable thickness - between 0.30 and 0.50 m - and of a greyish brown colour.

Planar Cross-bedded Sand Facies (Sp)

This facies includes medium to coarse unconsolidated sands, with low-angle planar cross-bedding. Beds are greyish, lenticular, with net or undulated contacts, with a thickness between 0.20 m and 0.70 m.

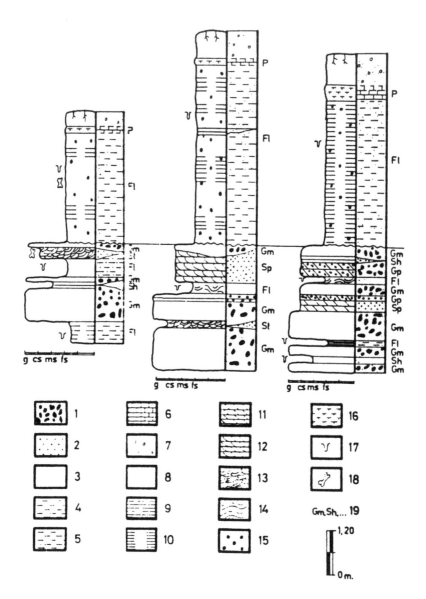

Figure 3. Detailed stratigraphic sequences of the Lower Psephitic Member of the Agua Blanca Formation in Bajo San José, Buenos Aires Province. 1) conglomerate, 2) coarse sand, 3) medium sand, 4) silty sand, 5) clayey silt, 6) carbonate, 7) soil, 8) massive bedding, 9) plane lamination, 10) poorly defined plane lamination, 11) planar cross-bedding, 12) tangential cross-bedding, 13) trough cross-bedding, 14) convolute bedding, 15) scattered pebble, 16) pedogenic features, 17) bioturbation, 18) fosile, 19) Gm, Sh, ... letter code devised by Miall (1978) for facies nomenclature.

Horizontal Bedding Sand Facies (Sh)

This facies includes massive or well/poorly defined parallel laminated medium to coarse sands, with scattered quartzite pebbles and pelitic intraclasts. In some sections, this facies presents typical paired horizontal laminations which are graded from coarse to silt, where each pair represents the deposition occurred during a single discharge cycle (Smith, 1974). Beds are greyish lenticular, with net or undulated contacts and a variable thickness -between 0.20 to 1.10 m.

3 PELITIC FACIES

Finely laminated Pelitic Facies (F1)

This facies includes very fine sands and sandy or clayey silts, with well or poorly defined horizontal fine bedding, ripple bedding, climbing ripples or massive bedding. Convolute bedding and pebble pockets are frequent. The consolidated sediments present a high degree of bioturbation with voids left by oxidated roots, numerous tiny channels and mottled structure. Beds are reddish to greenish, with net and undulated base and erosive top, and a thickness between 0.15 m and 5.0 m.

PALAEOENVIRONMENTAL INTERPRETATION

The sediments of the studied unit suggest frequent and extreme changes in flow conditions. Lateral and vertical variations are observed in grain size and sedimentary structures distribution which reflect velocity and depth fluctuations of the stream.

The general order of the sequence is fining-upward with cyclic sedimentation due to the transient nature of the flow with highly variable and extreme detritus discharge.

Facies Gm are interpreted as longitudinal bar deposits, built up by vertical accretion under the conditions of high flow regime, where clasts are transported by tracting flow and bedding development is not favoured.

Facies Gp appear associated with facies GM. They are the result of accretion processes in the distal extreme or lateral accretion due to the migration of linguoid bars during flooding periods. They may be also due to a delta-like growth from a remaining bar which was eroded in a relatively deep channel.

Both Facies St and Sp correspond to the migration of megaripples on the channel bed or on the bar surface.

Coarse psammites of Facies Sh would have been deposited under high fluidity conditions in plane bed phase of high flow regime. These sediments are due to the action of tractive flows on the channel bed and are part of the bars.

The horizontally laminated fine sands correspond to Facies Sh of low flow regime in shallow waters where vertical accretion processes take place.

Facies Fl represent sedimentation in inactive areas with scarce to abundant vegetation. Such areas are covered by water during flooding periods when flow speed is reduced by low depth and vegetation. This results in fine sediment trapping which generates stream structures of small scale. The set is affected by intense bioturbation and abundant root growth.

In accordance with the identified facies and with those established by Miall (1977), these sediments belong to bar deposits and channels developed from a braided fluvial system.

At present, this type of braided deposits characterize arid or semiarid regions or, in cases, paraglacial environments under extreme climatic conditions, scarce vegetation cover, unprotected slopes and with a big supply of coarse debris (Rust and Koster, 1984).

GEOLOGICAL HISTORY AND SEDIMENTARY IMPLICATIONS

In Figure 2, the stratigraphic scheme shows the lithostratigraphical units identified in the studied area. These units are equivalent to those described by Rabassa (1985) in the upper valley of Sauce Grande River. Two low fluvial terraces can be observed (Figure 2): a) Terrace I, composed by gravel and sands from the Lower Psephitic Member of the Agua Blanca Formation. In accordance with the identified fossil vertebrates and radiocarbon dating of snails (Rabassa, 1985; Figini et al., 1985; Borromei, 1985) this terrace has been assigned to a tentative Late Pleistocene age; b) Terrace II, composed by silts and sands of the Agua Blanca Formation Middle Sandy Member and Upper Silty-Sandy Member, by the sandy or clayey silts of the Chacra La Blanqueada Formation and by the eolian fine sands and silts of the Matadero Saldungaray Formation. According to the position in the stratigraphic sequence and correlation with those described by Rabassa (1985), both formations have been assigned to a Late Pleistocene-Holocene age.

The sediments of Terrace I overlie discordantly the Plio-Pleistocene regional substratum of the Saldungaray and La Toma Formations (Furque, 1973). The base of Terrace II has not been observed yet.

Terraces I and II are the geomorphic evidence of erosive processes associated with deep climatic changes and/or to relative lowering and rising of the sea level which affected directly the baselevel of the river. It must be taken into account the possible effects derived from tectonic activity due to NW-SE and E-W trending faults that have been observed in the studied area. E-W faults would probably control the present direction of the lower valley of the river (Bonorino et al., 1987), although the influence of the Late Pleistocene-Holocene E-W trending dune ridges in this area should not be neglected.

The sediments of the Lower Psephitic Member of the Agua Blanca Formation, as it has been previously demonstrated, correspond to a braided fluvial deposit which was probably generated as a consequence of a process of erosive rejuvenation in the basin of the Sauce Grande River, probably associated to a relative sea-level lowering (Rabassa, 1982; Vega et al., 1987) with a climatic deterioration during Late Pleistocene.

The change of baselevel could have triggered the rejuvenation of the drainage system (Schumm, 1977). The incision of the channel could have been initiated first at the mouth and then progressed upstream, rejuvenating the tributaries so that the main channel transported increasing quantities of sediments.

The recessing erosion would have been supported by a semiarid to arid and cold climate

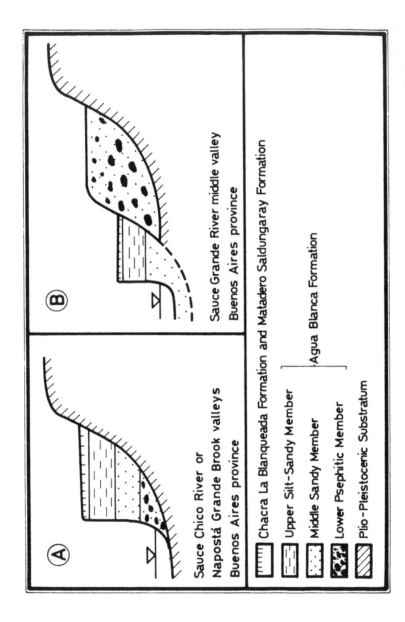

Figure 4. A comparison between the relative position of Late Pleistocene-Holocene fluvial deposits in (4a) the smaller basins of the area and (4b) the middle valley of the Sauce Grande River.

(Tricart, 1973; Tonni and Fidalgo, 1978) with scarce vegetation cover and unprotected slopes (Deschamps and Borromei, 1988).

As a result of the increasing sediment load, channels and bars were formed and the valley was widened by lateral erosion. However, as the tributaries and the main channel were adjusted to the new baselevel the sediment load decreased. The river that, at the moment, occupied the entire valley was transformed into a better-defined channel with active areas and inactive areas with vegetation and soil development. Meanwhile, the new incision of the valley continued, developing a low alluvial terrace (Terrace I), probably due to decreasing sediment yield (Schumm, 1977) or to tectonic activity (Furque, 1974).

Towards the end of the Late Pleistocene, the river, adjusting itself to the new baselevel, would have deposited the sediments of the Middle Sandy Member of the Agua Blanca Formation, perhaps under arid to semiarid climatic conditions with vegetation of halophilous steppe as it is inferred from equivalent sediments in the middle valley of Napostá Grande Creek (Quattrocchio et al., 1988).

The deposition of the Upper Sandy Silt Member of the Agua Blanca Formation would be related to a relative rise of sea level (Vega et al., 1987) and a climatic improvement by the development of paleosoils with vegetation of grass steppe and a decreasing discharge (Quattrocchio et al., 1988).

This unit is partially eroded with beheaded paleosoils probably by a climatic deterioration and it is covered discordantly by overbank facies of floodplain and/or eolian sediments from Chacra La Blanqueada and Matadero Saldungaray, respectively.

Meanwhile, during the Late Pleistocene and Holocene, the eolian sediments, silty sands and silt of the Saavedra Formation were deposited in the watersheds originated by the fluvial incision in the sediments of the Plio-Pleistocene substratum (Rabassa, 1982).

The middle valley of the Sauce Grande River (Figure 4b) presents geological and geomorphic characteristics which differentiate it from other valleys of the region (Figure 4a), whose headwaters are located in the Sierras Australes. A fining-upward fluvial sequence from gravels to sands and sandy silts can be observed in the banks of these rivers. These sediments correspond to the three members of the Agua Blanca Formation respectively.

ACKNOWLEDGEMENTS

To the Comisión de Investigaciones Científicas de la Provincia de Buenos Aires (C.I.C) for the financial support of this work.

REFERENCES

Bonorino, A.; Schillizzi, R. & Kostadinoff, J. 1987. Investigación geológica y geofísica en la región de Bahía Blanca. III Jornadas Pampeanas de Ciencias Naturales, Actas, 3:55-64.

Borromei, A. M. 1985. Sedimentos fluviales pleistocenos portadores de restos fósiles en el Bajo San José, Río Sauce Grande, Prov. de Buenos Aires. I Jornadas Geológicas Bonaerenses, Tandil, Abstracts.

Deschamps, C. & Borromei, A.M. 1988. Datos paleoambientales de la Formación Agua Blanca Miembro Psefítico Inferior (Pleistoceno Tardío?), Bajo San José (Prov. de Buenos Aires) en base a sus vertebrados. V Jornadas Paleontológicas de Vertebrados, La Plata. Abstracts.

Figini, A. & Furque, G. 1973. Descripción Geológica de la Hoja 34n Sierra de Pillahuinco. Serv. Nac. Min. y Geol., Boletín 141:70. Buenos Aires.

------. 1974. Sistema de terrazas aluviales en los aledaños de Saldungaray (Prov. de Buenos Aires),LEMIT. Serie II, No 265. La Plata.

Miall, A. 1977. A review of the braided-river depositional environment. Earth-Sci. Rev., 13:1-62.

------. 1978b. Lithofacies and vertical profile models in braided river deposits: a summary. In Miall, A. (ed.),Fluvial Sedimentology. Can. Soc. Petrol. Geol., Mem. 5, 597-604.

------. 1982. Analysis of fluvial depositional system.Am. Ass. Petrol. Geol., Calgary, Canada.

Quattrocchio, M.; Deschamps, C.; Martínez, D.; Grill, S. & Zavala, C. 1988. Caracterización paleontológica y paleo-ambiental de sedimentos cuaternarios, Arroyo Napostá Grande, Prov. de Buenos Aires.II Jornadas Geológicas Bonaerenses, Bahía Blanca.Actas: 37-46.

Rabassa, J. 1982. Variación regional y significado geomorfológico de la densidad de drenaje en la cuenca del río Sauce Grande, Prov. de Buenos Aires.Asoc.Geol.Arg., Rev. 37 (3):268-284.

------. 1985. Geología de los depósitos del Pleistoceno Superior y Holoceno en las cabeceras del Río Sauce Grande, Prov. de Buenos Aires.I Jornadas Geológicas Bonaerenses, Tandil. Abstracts.

Reineck, H. & Singh, I. 1980. Depositional sedimentary environments with reference to terrigenous clastics. Springer-Verlag. New York.

Rust, B. & Koster, B. 1984. Coarse alluvial deposits. In Walker, R. (ed.),Facies models, Geos. Can., Series 1, 53-69.

Schumm, S. 1977. The fluvial system. Wiley-Interscience, New York.

Smith, N. 1970. The braided stream depositional environment: comparison of the Platte River with some Silurian clastic rocks, North-Central Appalachians.Geol. Soc. Am. Bul., Vol. 81:2993-3014.

------. 1974. Sedimentology and bar formation in the Upper Kicking Horse River, a braided outwash stream.Journ. Geol., Vol. 82:205-223.

Spalletti, L. 1980. Paleoambientes Sedimentarios en Secuencias Silicocláticas.Asoc. Geol. Arg., Serie B, No 8. Buenos Aires.

Tonni, E. & Fidalgo, F. 1978. Consideraciones sobre los cambios climáticos durante el Pleistoceno Tardio-Reciente en la Provincia de Buenos Aires. Aspectos ecológicos y zoogeográficos relacionados. Ameghiniana XV (1-2):235-253. Buenos Aires.

Tricart, J. 1973. Geomorfología de la Pampa Deprimida.INTA. Col. Cient. XI. Buenos Aires.

Vega, V.; Valente, M. & Rodríguez, S. 1987. Ambiente marino somero y fluvial de los depósitos cuaternarios en las playas de Pehuen-Có, Buenos Aires, República Argentina.Reunión Final Proyecto N o 201 (IGCP-UNESCO) "Cuaternario de América del Sur". Ushuaia, Tierra del Fuego. Resúmenes.

H.F.FILIZOLA & L.COLTRINARI

Department of Geography, University of São Paulo, São Paulo, Brazil

11

Some comments on grain size indices of surficial deposits from a humid tropical environment: Taubaté basin, Brazil

ABSTRACT

Sedimentological parameters (mean size, skewness, standard deviation and kurtosis) were used to detect relations between grain size and selection degree of the mineral particles which constitute surficial formations in the Santa Catarina fluvial basin. Interrelations among the above mentioned parameters were also investigated in search of information related to the origin, process of transportation, and depositional environment of the examined materials.

The sample area is located in the upper-middle portion of Taubaté basin, a tectonic depression in the State of São Paulo (Brazil), and it is related to a remnant of São Jose dos Campos planation surface, probably of Plio-Pleistocenic age.

Results show very poorly selected materials which apparently point to depositional environments with variable energy. According to other authors, these characteristics relate to those of the materials; however, recent research suggests other possible interpretations.

RESUMEN

Los parámetros sedimentológicos (tamaño medio, skewness, desviación standard y kurtosis) han sido utilizados para detectar relaciones entre tamaño de grano y grado de selección de las partículas minerales que constituyen formaciones superficiales en la cuenca fluvial de Santa Catarina. Las interrelaciones entre los parámetros mencionados más arriba fueron también investigados en la búsqueda de información relacionada al origen, proceso de transporte y ambiente depositacional de los materiales examinados.

El área de muestreo está ubicada en la porción media superior de la cuenca de Taubaté, una depresión tectónica en el Estado de São Paulo (Brasil), y está vinculada a un remanente de la superficie de planación de São Jose dos Campos, probablemente de edad Plio-Pleistocena.

Los resultados muestran materiales muy pobremente seleccionados los cuales apuntan aparentemente a ambientes depositacionales con energía variable. De acuerdo a otros autores, estas características se relacionan con aquellas de los materiales; sin embargo, investigaciones recientes sugieren otras posibles interpretaciones.

INTRODUCTION

The use of data related to surficial formations in geomorphological studies is very recent, especially when considered as relevant as the landforms themselves. It should not be seen as a mere incorporation of a research technology but as an enrichment of tridimensional geomorphological analysis with physical and geochemical data resulting from distinct palaeoenvironmental processes.

According to Faure (1978), surficial formations play a fundamental role in geomorphology and Quaternary geology, since they are located at the earth's surface; thus, they are geological formations in close contact with biosphere and atmosphere. Reworked by all the mechanisms involved in the evolution of landforms on which they are momentarily resting upon surficial sediments help to understand the surficial dynamics during recent geological times.

As a consequence of biogeochemical transformations provoked by distinct environmental conditions, these materials changed their mechanical and chemical characteristics; due to these changes, the behaviour of surficial formations towards morphogenesis has been modified as well (Coltrinari, 1987). This paper intends to contribute to the knowledge of surficial formations considered as correlative deposits of external geodynamics during the latter stages of landscape evolution and consequently, a source of palaeoenvironmental data.

This type of research is relevant also when the importance given to stratigraphic studies in modern Quaternary research is considered (Bowen, 1978), since they supply information for the reconstruction of past climates and environments by means of their biotic and abiotic contents.

In southeastern Brazil, the humid tropical climate was unfavourable for the permanence of adequate materials for bio- and chronostratigraphic determinations; therefore, research is often confined to the use of more classical techniques, allowing at least the preparation of lithostratigraphical columns. These sequences include surficial formations associated to different topographic levels or geomorphological units, and help to identify their sedimentological features, as shown in this paper. Considering the lack of more reliable indicators, these data are the starting point for the proposal of hypotheses concerning the sequence of paleoclimatic episodes and the recent evolution of São Jose dos Campos plateau.

When examining sediments from the sample area (Figure 1), the parameters proposed by Folk & Ward (1957) were used. They emphasized the importance of mean grain size and sorting of materials to infer transport and deposition modes of sediments. Though these parameters are not very well known and rather poorly understood, they have been considered useful.

Coltrinari & Filizola (1986) have published results concerning statistical grain size indixes of the study area and Coltrinari (1987) employed those data to identify surficial formations of interfluves and slopes; they did not discuss their meaning or value.

THE SAMPLE AREA

Santa Catarina plateau is located on the right margin of Paraíba do Sul river (Figure 1) about 10 km from São Jose dos Campos. It forms part of São Jose dos Campos plateau, actually dissected by the right bank tributaries of the main river which come from the NW slope of the Serra do Mar. Sampling of surficial formations was undertaken in an area of approximately 20 sq. km.

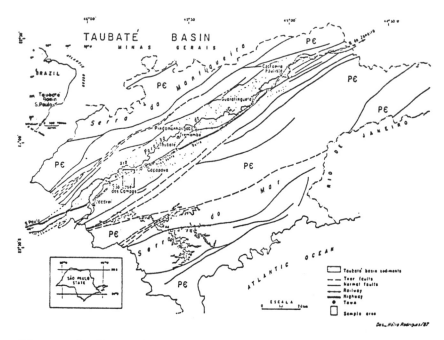

Figure 1. Location Map of The Sample Area.

GEOLOGY

Landforms in the studied area were eroded on
Caçapava Formation sedimentary rocks (Carneiro et
al., 1976), the upper unit of the Taubaté Group
reaching up to 520 m of thickness in the Taubaté
tectonic basin. The origin and evolution of the
Taubaté graben are associated to those of the
continental margin beginning with the Wealdenian
Reactivation (Almeida, 1976). Normal faulting
occurred along old Precambrian fault-lines causing
gravitational block-sliding and tilting; the Serras
do Mar and da Mantiqueira escarpments are
conspicuous examples of these crustal processes.

The western subsidized part of the Serra do
Mar block together with the Mantiqueira fault line
escarpment gave origin to the narrow basin occupied
by the Taubaté Group. According to Mello et al
(1985) three types of Tertiary sedimentary
facies can be found; one of them is predominantly
lacustrine and its outcrops appear only along the
basin axis, around the city of Taubaté (Tremembé
Formation). The second one is a conglomerate
facies, constituted by breccias and polymictic
conglomerates with sandy and silty matrix and it is
found mainly along the NW border of the basin; its
origin is associated with coalescence of alluvial
fans near the tectonic escarpments. The last one is
a fluvial facies (Caçapava Formation) and contains
sediments of almost all grain sizes showing
remarkable textural and mineralogical immaturity.
It is found in all outcrops in the research area
between Jacareí and São José dos Campos.

According to Carneiro et al. (1976), Caçapava
Formation is predominantly sandy and silty, though
coarser beds occur in the sample area. Vertical and
lateral lithological variations are frequent.

MORPHOLOGY

Caçapava Formation sedimentary rocks are associated
with the landforms of São José dos Campos plateau;
according to Ab´Saber (1969), the highest hills
would be the remnants of the highest level of São
José dos Campos planation surface of Plio-
Pleistocenic age (615-650 m of altitude). At
present, the highest interfluves reach up to 700 m
along the SE border near Serra do Mar.

A detailed survey of surficial formations was
undertaken in the Santa Catarina plateau, bordered
by the Porangaba and da Divisa rivers valleys
(Figure 2). Western and southwestern margins show
long complex slopes along which the Porangaba
stream and its tributaries run (565 m). Those
tributaries as well as those of da Divisa stream
dissected the lower part of the slopes; their
sources occur at 600 m.

The highest altitudes in Santa Catarina
plateau are found on the SE/E angle, around 650-680
m; landforms and surficial formations differ
clearly from those on the second topographic step,
formed by hill-interfluves between 625-645 m, which
in the northern part reach 620 m. Here, the

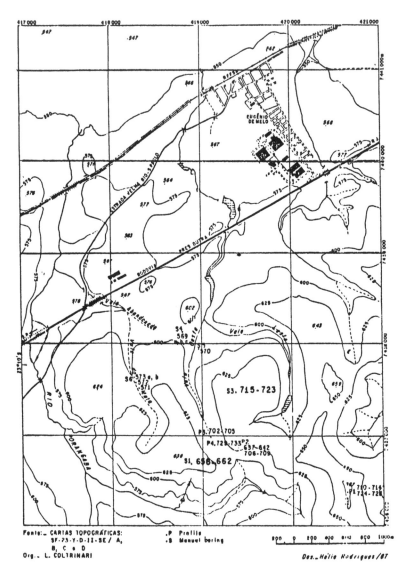

Figure 2. Surficial formations - Santa Catarina Basin.

landforms are less clearly dissected, partly because thick sandy-clayey formations cover the contact between the hills and the alluvial levels; the latter are found between Presidente Dutra highway and the railway (Coltrinari, 1983) (Figure 2).

The valley bottoms of the Santa Catarina fluvial system occur between 585-610 m, and the alluvial levels of Paraíba do Sul terraces between 565-575 m. The lowest terrace is found towards 560 m and the base level formed by the Paraíba do Sul floodplain at 540-550 m (Coltrinari, 1983).

SURFICIAL FORMATIONS AND THEIR ATTRIBUTES

1 LOCATION OF SAMPLES

According to field criteria, surficial formations were divided in interfluve and slope formations, and terraces and valley bottom formations (Figure 2).

In the first group, profiles P1 and P2, and manual borings S1, S2 and S3 are included. Materials sampled in P1 (710-714 and 724-728) are located between 675-680 m. Profile P2 (samples 637-642 and 706-709) and manual borings S1 (656-662) and S3 (715-723) were described between 630-635 m; they are associated to interfluves and higher slopes in the second topographic step.

In the second group, profiles P3 (702-705) and P4 (729-733), manual borings S4 (569 a, b, c), S5 (570) and S6 (573 a, b) and loose materials from destroyed ant-hills (571 and 572) found between 585-615 m, have been included.

2 STATISTICAL PARAMETERS

To determine physical features of sediments and gather information about their hypothetical origin mean size (Mz), inclusive graphic standard deviation (6I), inclusive graphic skewness (SkI), and graphic kurtosis (Kg) were used (Figure 3 & 4) (Folk & Ward, 1957).

	Samples	GRAIN-SIZE PARAMETERS (1)		Sk_I	K_G
		M_2	δI		
P1	710	5,53	2,77	-0,42	0,99
	711	5,83	2,40	-0,33	1,06
	712	5,83	2,60	-0,39	1,04
	713	5,97	2,49	-0,07	1,08
	714	4,73	2,14	-0,19	1,42
	724	5,85	2,96	-0,67	0,66
	725	5,40	3,42	-0,72	0,78
	726	5,93	3,09	-0,62	0,77
	727	5,45	3,61	-0,70	0,83
	728	5,77	3,02	-0,63	0,76
P2	637	4,13	3,08	-0,01	0,65
	638	3,90	3,02	-0,04	0,53
	639	4,77	3,18	-0,27	0,63
	640	4,47	3,31	-0,31	0,66
	641	4,73	3,33	-0,44	0,71
	642	5,13	3,33	-0,42	0,76
	706	4,27	2,86	0,01	0,71
	707	1,03	3,63	0,69	0,46
	708	2,77	3,34	-0,01	0,60
	709	1,53	3,72	0,41	0,46
S1	715	5,53	2,77	-0,63	0,75
	716	5,47	2,90	-0,62	0,69
	717	5,53	2,94	-0,62	0,70
	718	5,5	2,96	-0,70	0,67
	719	5,43	2,88	-0,78	0,69
	720	5,53	2,90	-0,69	0,67
	721	5,70	2,90	-0,64	0,67
	722	5,63	2,93	-0,675	0,69
	723	5,53	3,11	-0,60	0,63

Figure 3. Grain-size parameters (I).

3 MEAN SIZE (Mz)

Inman (1952) proposed the following equation

$$M \emptyset = \frac{1}{2} (\emptyset\ 16 + \emptyset\ 84)$$

to calculate mean grain size. This equation fits nearly normal curves but it does not reflect accurately the mean size of bimodal and strongly skewed curves. Therefore, Folk & Ward (1957) used another measure of the mean (Mz) determined by the formula

$$Mz = \frac{\emptyset\ 16 + \emptyset 50 + \emptyset\ 84,}{3}$$

where the $\emptyset$ 16 may be considered roughly as the median of the coarsest third, and the $\emptyset$ 84 as the average size of the finest third; the average of the middle third completes the picture giving a better representation of the true phi mean. According to Suguio (1973), Mz reflects the general mean of sediments size being influenced by the supply source of materials, by deposition process and transportation velocity.

4 INCLUSIVE GRAPHIC STANDARD DEVIATION (σI)

As a measure of sorting, Inman (1952) followed other authors and suggested phi standard deviation,

$$\sigma\emptyset = \frac{\emptyset 84\ -\emptyset\ 16;}{2}$$

which is adequate for many normal curves, but it is based only on the central part of the distribution ignoring the "tails", in which some of the most valuable information is found. Therefore, Folk & Ward (1957) included more of the distribution curve in the sorting measure, called Inclusive Graphic Standard Deviation, determined by the equation,

$$\sigma I = \frac{\emptyset 84 - \emptyset 16}{4} + \frac{\emptyset\ 95 - \emptyset\ 5}{6.6}$$

They also suggested the following divisional points: σI under 0.35, very well sorted; σI 0.35 - 0.050, well sorted; σI0.50-1.00 moderately

	Samples	Mz	σ_{I2}	Sk_I	K_{G1}
S3	656	4,93	3,02	-0,34	0,71
	657	4,43	2,90	-0,16	0,64
	658	4,43	2,88	-0,15	0,63
	659	4,50	3,03	-0,09	0,65
	660	4,53	3,11	-0,19	0,71
	661	4,63	2,95	-0,28	0,63
	662	4,73	3,06	-0,28	0,66
P3	702	3,9	2,94	0,17	0,58
	703	3,6	3,14	0,26	0,61
	704	3,8	2,71	0,22	0,56
	705	3,7	2,93	0,24	0,60
P4	729	3,58	3,38	0,485	0,58
	730	3,32	2,94	0,41	0,59
	731	3,53	3,03	0,36	0,58
	732	3,48	3,0	0,32	0,58
	733	3,73	2,85	0,30	0,58
S4	569a	3,23	3,04	0,53	0,57
	569b	3,92	2,90	0,25	0,55
	569c	3,63	3,09	0,31	0,57
S5	570	3,0	3,25	0,65	0,56
S6	573a	2,83	2,99	0,61	0,52
	573b	3,13	3,09	0,52	0,52
Loose sediments	571	2,93	3,18	0,63	0,54
	572	4,57	2,91	-0,39	0,54

GRAIN-SIZE PARAMETERS (II)

Figure 4. Grain-size parameters (II).

sorted; δI 1.00-2.00, poorly sorted; δI 2.00 - 4.00, very poorly sorted, and over 4.00, extremely poorly sorted. About standard deviation meaning, Suguio (1973) considered that the sorting of sediments depends, to a certain degree, on their grain size; results are more precise for sandy and coarser materials, being poorer when finer sediments are concerned.

5 INCLUSIVE GRAPHIC SKEWNESS (SkI)

To measure the overall skewness, Folk & Ward (1957) suggested the formula

$$SkI = \frac{\phi\ 16 + \phi\ 84 - 2\ \phi\ 50}{2\ (\phi\ 84 - \phi\ 16)} + \frac{\phi\ 5 + \phi\ 95 - 2\ \phi\ 50}{2\ (\phi\ 95 - \phi\ 5)}$$

By using this equation, skewness is geometrically independent of sorting, perfectly symmetrical curves have SkI = 0.00, and the absolute mathematical limits are -1.00 to +1.00; however, very few curves have SkI beyond -0.80 and +0.80. Positive values indicate that samples have a "tail" of fines; negative values indicate a tail of coarser grains. In bimodal, samples asymmetry is small when the two modes are approximately equivalent; however, when the differences are wide, the degree of asymmetry increases and the signal (positive or negative) is determined by the relationship between the modes.

6 GRAPHIC KURTOSIS (Kg)

The Graphic Kurtosis used by Folk & Ward (1957) is given by the equation

$$Kg = \frac{\phi\ 95 - \phi\ 5}{2.44\ (\phi\ 75 - \phi\ 25)}$$

For normal curves, Kg = 1.00; in very platykurtic or deficiently peaked curves, Kg under 0.67; Kg = 0.67 - 0.90, pltykurtic; kg = 0.90 - 1.11, mesokurtic; Kg = 1.11 - 1.50, leptokurtic; Kg 1.50 - 3.00, very leptokurtic, and Kg = over 3.00, extremely leptokurtic. A curve with Kg = 2.00 could represent a relatively well sorted sediment in the central part of the distribution.
    Folk & Ward (1957) considered that Kurtosis

measures the ratio of the sorting in the extremes of the distribution compared with the sorting in the central part and, as such, it is a sensitive and valuable test of the normality of a distribution. Together with asymmetry kurtosis, it is thought to be a rather good parameter to differentiate sedimentary environments; McLaren (1981) wrote that kurtosis does not provide further information for the interpretation of a grain-size distribution. Actually, its meaning is rather obscure and data about magnitude and frequency of kurtosis are yet scarce.

STATISTICAL PARAMETERS OF SURFICIAL FORMATIONS IN THE STUDIED AREA

Samples from Santa Catarina plateau show mean size between Ø= 5.97 (coarse silt) and Ø= 1.03 (medium sand); none of the cumulative frequency curves correspond to a normal distribution. In profile P2, samples 706 and 708 - which correspond to fine grains mixed with the "cascalheira" (gravel layer) - are the ones whose curves are nearer to the normal curve. The widest withdrawal from normality occurs in sediments beyond Ø 6.0 (below 0.01 mm) corresponding to silt and clay.

Samples 707 and 709, from the gravel bed in the lower part of P2, present the greatest withdrawal from the normal in the entire area. It could be said that gravels were deposited in an environment whose energy was greater than that of finer materials, probably deposited by infiltration during a later stage of sedimentation.

1 STATISTICAL PARAMETERS AND POSITION OF SAMPLES IN LANDSCAPE

Materials sampled between 660-680 m (P1) are homogeneous either in relation to Mz or to 6 I (Figure 3). Mean size ranged from Ø= 4.73 and 5.93 corresponding to very coarse silt/coarse silt sizes. Standard deviation indicates very poorly selected grains (2.15-3.61); asymmetry limits are -0.07 and -0.72 (nearly symmetrical to very negative-skewed), and kurtosis values fall between 0.66 and 1.42 (very platykurtic to leptokurtic). The lower beds (724-728) show platykurtic distribution, except for sample 724,

with "saddle distribution" due to two equivalent
and widely separated modes. In the upper set (710-
714), there is only one sample - 714 -
corresponding to a leptokurtic distribution, the
other materials are mesokurtic, with Kg indices
near to 1.0.

Hills in the lower topographic unit are
predominantly covered by fine sediments, with the
exception of beds 706-709, in the lower part of P2
(Figure 2). These samples are heterogeneous in
relation to mean grain size in Santa Catarina
plateau samples; the upper set Mz values range from
Ø 3.9 (very fine sand to coarse silt); sediments
associated to the "cascalheira" have as Mz limits
Ø 1.03 (medium sand) and 4.27 (very coarse silt),
with gravels beyond Ø 3.3 (10 mm). All ten samples
in P2 are very poorly sorted (б I 2.86-3.72).
Asymmetry varies from very negative to very
positive. In the coarser beds (707, 709), tails are
formed by finer materials (positively
asymmetrical). Samples 706 and 708 are
approximately symmetrical. The upper beds of P2
show negative asymmetry or are nearly symmetrical;
Kg values are those of platykurtic and very
platykurtic distributions.

On a similar topographic level as P2, samples
were collected from S1 and S3. All parameters
considered (Figure 3), S3 appears as the most
homogeneous: Mz Ø 5.43 to 5.70 (coarse silt), б I
corresponding to poorly selected materials (2.77-
3.11). Asymmetry is very negative and platykurtic
curves predominate; sample 723 is very platykurtic.
In S1 data are similar to those in S3.

Sediments collected in valley bottoms and
associated deposits (Figure 4) may also be
considered as presenting a remarkable degree of
similarity among themselves. Mean grain-size ranges
between Ø 2.83-4.57 (fine sand to very coarse
silt); sorting is poor (б I 2.9-3.4); asymmetry is
positive to very positive, excepting sample 572;
and finally, kurtosis shows very platykurtic
curves.

2 INTERRELATIONSHIPS OF THE SIZE PARAMETERS

Folk & Ward (1957) plotted the four size parameters
against each other as scatter diagrams. They
searched for a better understanding of the
geological significance of size parameters and

hoped that it would be possible to add new criteria for the identification of environments by size analysis.

Figures 5 to 10 show two-variable scatter plots, as in Folk & Ward (1957), but in this paper they are only used to precise mathematical relationships among size parameters.

3 MEAN SIZE (Mz) VERSUS STANDARD DEVIATION (σI)

When a wide range of grain sizes (gravel to clay) is present in this type of plot, scatter bands frequently form some segment of a broadened M-shaped trend. Often, only a V-shaped or inverted V-shaped trend develops if the size range is smaller and only one limb of the V occurs if the size range is too small. In Figure 5, the absence of very coarse mean grain size determines that of the left limb of the M. Data scattering shows that finer sediments group themselves along the right limb of the M; highest values (poorest sorting) related to medium sand appear on the left and those corresponding to medium silt, on the right. A relative improved sorting is observed in samples with very fine sand mean-size; on the contrary, sorting is worse in finer sediments (p2, S1, S3). Actually, these samples show almost equal percentages of sand and clay.

It must be remarked that clayey materials from P1 are the least poor sorted samples in the area, and that coarse beds in P2 are the worst ones.

4 MEAN SIZE (Mz) VERSUS SKEWNESS (SkI)

According to Folk & Ward (1957), the pure modal fractions are in themselves nearly symmetrical but the mixing of two modes produces negative skewness if the finer mode is most abundant and positive skewness if the coarser mode is most abundant. A symmetrical curve results from an equal quantity of the two modes. In Figure 6 it can be clearly observed that the greatest part of samples is formed by sediments finer than Ø 2.5 Mz; around Ø 4.0, there is a clear partition between positive and negative skewness in the area. On the right side, under the normal curve, clayey and sandy-clayey materials from interfluves
and slopes appear. From left to right, the amount

248

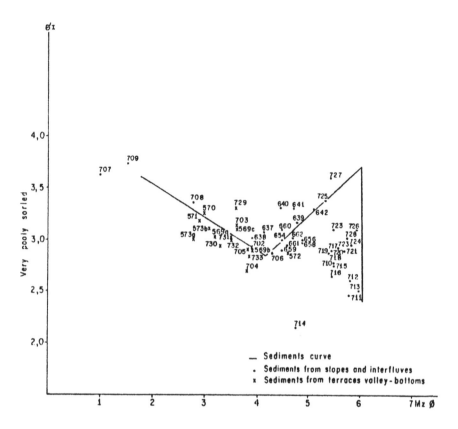

Org._ HELOISA F. FILIZOLA
Des._ Hélio Rodrigues/87

Figure 5. Mean size versus Standard deviation.

of clay mounts, and the size-curve become more and more negative-skewed.

On the left side, above the normal curve, sandy and sandy-clayey sediments (predominantly from valley bottoms) are found. Samples become more and more positive-skewed as the amount of sand comes to exceed the amount of clay and also when coarse sand percentages increase. This is the case for samples on the extreme left of the sediment curve beginning with 569 a.

Addition or substraction of a mode or modes have been interpreted as a function of depositional environments, i.e. as being purely

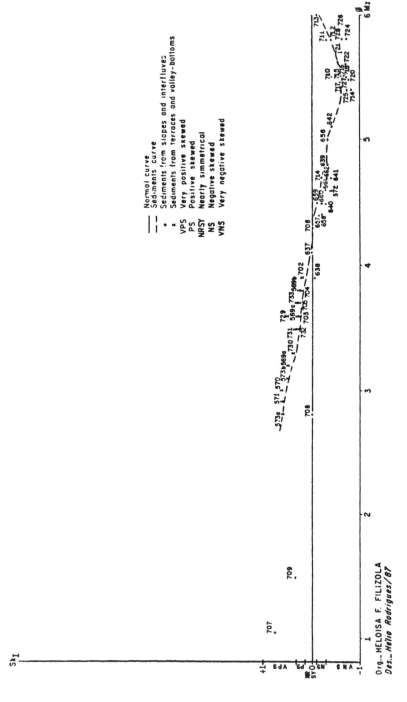

Figure 6. Mean size versus Skewness.

250

mechanical/physical in origin. In this case, it is possible to propose a different hypothesis to explain the growing negative asymmetry associated to the highest interfluves in the area; growing percentages of clay are associated with deeply weathered profiles where there are evidences of rather old pedogenetic episodes.

Evidences of geochemical processes are present on surfaces of quartz grains belonging to Profile 1; shapes of cavities are cavernous or flattened, sometimes almost divided by deep fissures. Similar features are also found in profile P2, in grains of fine materials.

5 MEAN SIZE (Mz) VERSUS KURTOSIS (Kg)

The scattered plot of kurtosis versus mean size (Figure 7) shows that the next to pure clay mode in samples from profile 1 gives nearly normal curves with Kg=1.00. Other materials are platykurtic or very platykurtic. With a few exceptions, all very platykurtic sediments are predominantly sandy and they were collected in valley bottoms. Two samples, 707 and 709, belong to the lower part of profile 2, and they are formed by the coarsest samples in the area.

6 STANDARD DEVIATION (6 I) VERSUS SKEWNESS (SkI)

Interrelations between both parameters in the sample area are shown in Figure 8. Folk & Ward (1957) wrote that: "... if standard deviation is a function of mean size, and if skewness is also a function of mean size, ...sorting and skewness will bear a mathematical relation to each other..." . Symmetrical curves can be associated with unimodal samples with good sorting or can be obtained from equal mixtures of the two modes which have the poorest sorting for a suite of samples.

In Santa Catarina materials, the scattered plot forms a trend with the shape of an orthogonal axis as compared with the normal curve. Extreme skewness values are found in samples with a dominant mode and a subordinated one. With all sediments in the area, very poorly sorted scattering is limited to a series of groups distributed on both sides of the normal curve. Only samples 667, 638, 706 and 708 remain near it with a

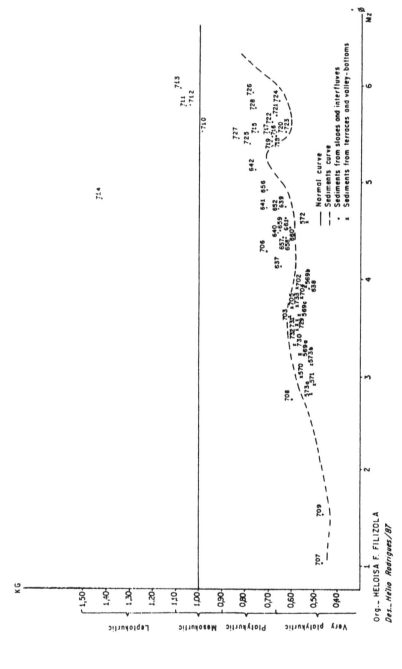

Figure 7. Mean size versus Kurtosis.

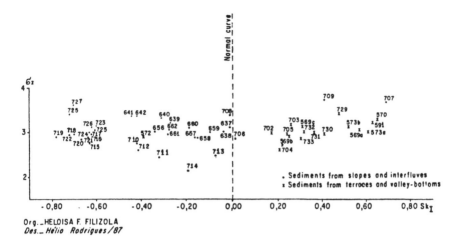

Figure 8. Standard deviation versus Skewness

dominant sand mode and quite approximate values of
silt and clay percentages. The other samples are
distributed according to the predominance of fine
and coarse sediments on the left and right sides of
the curve, respectively.

7 STANDARD DEVIATION (δI) VERSUS KURTOSIS (Kg)

Unimodal sediments produce normal kurtosis and are
best sorted (Figure 9); this is once more the
case with samples in profile P1. The other
sediments group themselves in a "cloud"
corresponding to low kurtosis values.

8 SKEWNESS (SkI) VERSUS KURTOSIS (Kg)

Folk & Ward (1957) believed that these two
properties depend on the proportions of the two
present modes. It will be noted that in the samples
studied only very few analyses fall in the range of
what would be considered "normal" curves; in fact,
we have three samples included within "normal"
values of kurtosis, and five within the limits of
"normal" skewness; none of the samples is normal
with regard to both skewness and kurtosis (Figure
10). Most of the sediments show great departures

253

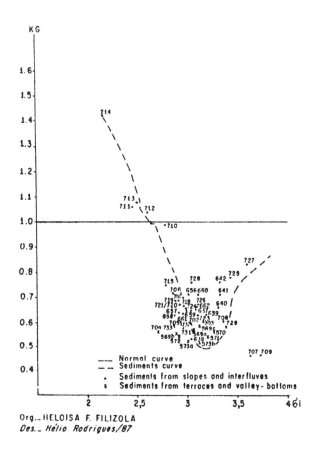

Figure 9. Standard deviation versus Kurtosis.

from normality; the distribution along an axis on the left side of the scattered plot is quite similar to that shown in Figure 8.

SOME COMMENTS ON DATA AND THEIR SIGNIFICANCE

Sedimentological parameters were used as a tool to search for reliable information concerning surficial formations in the humid tropical southeastern Brasil. Folk & Ward (1957) parameters - mean size, standard deviation, skewness and kurtosis - have been used to establish a tentative differentiation between sediments belonging to different geomorphological units and topographic surfaces.

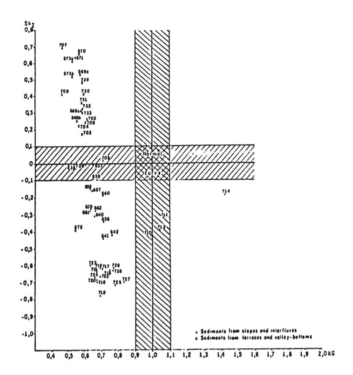

Figure 10. Kurtosis versus Skewness.

As mathematical relations are considered, analyzed data may be adequate, i.e. inasmuch they help to precise sedimentological attributes of surficial formations. On the other side, interpretation concerning transport and sedimentation dynamics should be discouraged at least in morphoclimatic zones where biogeochemical weathering has been -as it is today- a predominant process in landscape modelling.

REFERENCES

Ab´Sáber, A.N. 1969. O Quaternário na Bacia de Taubaté: estado atual dos conhecimentos: São Paulo, IGEOG/USP, 23pp. (Geomorfologia 7).
Almeida, F.F.M. de. 1976. The system of continental rifts bordering the Santos Basin, Brazil. In: International Symposium on Continental Margins of Atlantic Type, 1, São Paulo, 1975. An. Acad. Bras. Ci, 46(supl.), 15-26. São Paulo.

Bowen, D.O. 1978. Quaternary Geology-A stratigraphic framework for multidisciplinary work. 221pp. Great Britain:Pergamon.

Carneiro, C.D.R.; Hasui, Y. & Giancursi, F.D. 1976. Estrutura da Bacia de Taubaté na região de São José dos Campos. In: Congr. Bras. Geol. 29 Ouro Preto, 1976. Anais..., 4:247-256, il. SBG, Ouro Preto. 1976.

Coltrinari, L. 1983. O Cenozóico Superior da margem continental brasileira. In: Reunión del Grupo Español de Trabajo de Cuaternario, 6, Santiago/Vigo, 1983. VI Reunión..., O Castro/Sada. Cuadernos del Laboratorio Xeológico de Laxe, 5:565-593.

Coltrinari, L. 1987. Slope forms and deposits in Taubaté basin, southeastern Brazil. In: INQUA Congress, 12, Ottawa, 1987. Programme... p.147. INQUA. Ottawa.

Coltrinari, L. & Filizola, H. 1986. Grain size parameters of surficial formations of São José dos Campos plateau (SE Brazil). In: Reunión del Proyecto 201 Cuaternario de América del Sur - IGCP/UNESCO, 1. Mérida, dic. 1986. Resúmenes...Mérida, IGCP/UNESCO, 13-, 1986.

Faure, H. 1978. Continuité et descontinuité dans la génese des formations superficielles. In: Colóquio interdisciplinar Franco-Brasileiro "Estudo e cartografia das formações superficiais nas regiões tropicais e suas aplicações", São Paulo, 1, 1978. Comunicações...USP/FFLCH/IGEOG, 1:39-54. São Paulo.

Folk, R.L. & Ward, W.C. 1957. Brazos River bar: a study in the significance of grain size parameters. J. Sediment. Petrol., 27:3-27.

Inman, D.L. 1952. Measures for describing the size distribution of sediments. J. Sediment. Petrol., 22:125-145.

Mc Laren, P. 1981. An interpretation of trends in measures of grain size. J. Sediment. Petrol., 51(2):611-624.

Mello, M.S.; Riccomini, C.; Hasui, Y.; Almeida, F.F.M. de & Coimbra, A.M. 1985. Geologia e evolução do sistema de bacias tafrogênicas continentais do Sudeste do Brasil. Rev. Bras. Geoc., 15(3):193-201. São Paulo.

Suguio, K. 1973. Introdução a Sedimentologia. 317pp. São Paulo: E.Blucher/EDUSP.

OMAR R.LAPIDO
Dirección Nacional de Minería y Geología, Buenos Aires, Argentina

CARLOS A.BELTRAMONE & MIGUEL J.HALLER
Centro Nacional Patagónico-CONICET, Puerto Madryn, Argentina

12

Glacial deposits on the Patagonian Cordillera at latitude 43°30' S

ABSTRACT

Deposits of three glacial events were recognized in the area of the upper Río Tecka, a stream of the Atlantic drainage basin of the Patagonian Andes.

Till of the oldest glaciation is located at elevations of 1300-1600 m, hanging on the high shoulders of the valleys of the later glaciations and may correspond to a mountain ice sheet.

On the eastern slope of the Cordón Caquel a sharp erosion surface covered by pediment gravels is developed at an elevation of 1200 to 900 m, on the old glacial deposits and the underlying Tertiary sediments.

In the Río Tecka valley, an external terminal moraine at 850 m is correlated to the El Condor Glaciation. Two well preserved end moraine belts at an elevation of 750-850 m, are assigned to the Late Wisconsinan Nahuel Huapi glaciation.

RESUMEN

En la región del Alto Río Tecka, un curso de agua de la vertiente atlántica de los Andes Patagónicos, se han reconocido depósitos sedimentarios correspondientes a tres eventos glaciales.

Till de la glaciación más antigua se encuentra a elevaciones de 1300-1600 m s.n.m., colgante sobre las altas hombreras de los valles de las glaciaciones posteriores y podría corresponder a un manto de hielo de montaña.

Sobre las laderas orientales del Cordón Caquel

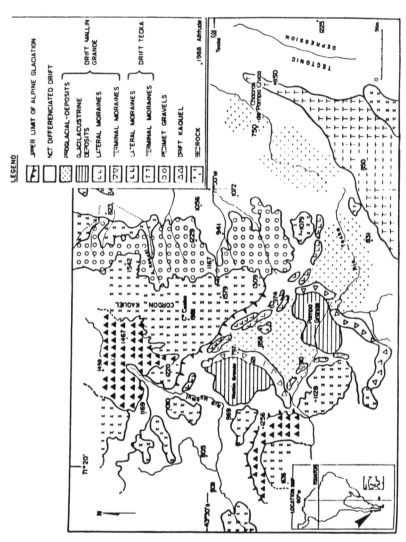

Figure 1. Glacial deposits on the Río Tecka valley, Patagonian Cordillera (43o30'S).

se ha desarrollado una abrupta superficie de erosión cubierta por gravas de pedimento, a elevaciones de 1200-900 m s.n.m., sobre los depósitos glaciales antiguos y los sedimentos terciarios infrayacentes.

En el valle del Río Tecka, una morena terminal externa que se encuentra a 850 m s.n.m., se ha correlacionado con la glaciación El Cóndor. Otros dos cinturones morénicos bien preservados, ubicados a elevaciones de 750-850 m, se asignan a la Glaciación Nahuel Huapi, del Pleistoceno Superior tardío (Late Wisconsinan).

INTRODUCTION

Previous workers on the glacial stratigraphy of the eastern slope of the Patagonian Andes report the presence of four (Caldenius, 1932) or three (Flint & Fidalgo, 1963; Rabassa, 1983) major glacial episodes during the Pleistocene, and recognized Holocene glacial deposits at different latitudes (e.g. Ramos, 1981 and Haller, 1981).

These stratigraphic schemes have been extended to the entire Patagonian Cordillera region. Detailed studies at several localities led to the development of local -mostly relative age dated- glacial chronologies and the recognized glacial records were correlated to the regional stratigraphic picture.

In this paper we describe the stratigraphy and characteristics of the Pleistocene (and Late Tertiary?) deposits of a transverse valley of the southern segment of the Northern Patagonian Andes, located at Latitude 43o 30´ South. We also propose a correlation of these deposits to other glacial stratigraphies developed for both sides of the Patagonian Cordillera. In addition, we include a model of alternating Quaternary climatic changes and the resulting morphogenesis in this segment of the Southern Andes.

GEOGRAPHICAL AND GEOLOGICAL SETTING

The Río Tecka is a cordilleran stream which flows in an east-west-trending, fluvially modified glacial trough (see Figure 1). This valley is about 30 km long with the towns of Corcovado near the eastern end. The continental water shed runs along

a transverse Pleistocene morainic arc located about
10 km east of Corcovado. In the east, the Río Tecka
curves to the north, following a tectonic
depression.

Field observations were also made along both
slopes of the Cordón Caquel, the easternmost
Cordilleran range, which extends from the Tecka
valley to the north. These studies were undertaken
to develop a better understanding of the
morphogenetic history of the area.

The bedrock in the study area consists of
Jurassic volcanoclastics, Cretaceous volcanics and
related shallow intrusives, and Miocene lacustrine
sediments. These rocks are tilted and fractured.

PLEISTOCENE DEPOSITS

Several deposits of glacial and non glacial origin
have been recognized at the Río Tecka valley. Their
relative chronology was established based on
topographic and geomorphic evidence:

Lithostratigraphic unit	Glacial event
Mallín Grande Drift	Nahuel Huapi
Tecka Drift	El Cóndor
"Pediment Gravels"	
Caquel Drift	Pichileufu

WESTERN SLOPE OF CORDON CAQUEL

CAQUEL DRIFT

On the western slope of the Cordón Caquel a sheet
of glacigenic deposits covers an extensive,
relatively smooth surface at elevations ranging
from 1300 to 1600 m - more than 500 m above the
current floor of the Río Tecka valley. This till
cover is almost continuous, with an average
thickness of 5-6 m and a maximum of 18 m. Small
patches of till located at the same altitude on the
eastern flank are also included in this unit. The

Caquel Drift is composed of massive till, with red-brownish silty-clay matrix and large (up to 2.5 m) granitic and andesitic boulders with polished faces. Some of these boulders are striated. At some locations striae were also observed on the underlying bedrock. Drainage is poorly developed, with numerous ponds and the streams hanging on the main valley. The surrounding ranges consist mostly of domelike hills and rounded ridges with gentle slopes and a few sharp peaks such as Cerro Cuche (1988 m) which are interpreted as ancient nunataks. The tills are interpreted as ground moraine deposited by an ice sheet which extended eastward from the Cordilleran core. The outwash deposits were probably located far to the east, but they have been removed or at least greatly modified by later fluvial processes.

It was not possible to obtain limiting ages for the Caquel Drift. Based on its topographical position and geomorphological relationships with landforms and deposits of younger glaciations, we assume that the Caquel Drift is a remnant of an Early Pleistocene (?) event that affected the area.

EASTERN SLOPE OF CORDON CAQUEL

"PLEISTOCENE PEDIMENT GRAVELS"

This informal name is given to the gravel blanket which unconformably overlies Cretaceous volcanics, Miocene sedimentary rocks and till of the Caquel Drift on the eastern flank of Cordon Caquel at altitudes from 900 to 1200 m. The gravel cap is locally well preserved, consisting of rounded and polished pebbles of andesites in a matrix of silty sand, poorly indurated. The thickness averages 3-4 m, but reaches 10 m at some places. The pediment was graded to an older, higher level of the floor of the tectonic depression to the east. This surface is the highest and westernmost of a series of dissected pediment levels, which cover much of eastern Patagonia and are traceable almost without interruption to the Atlantic Ocean.

Studies indicate that development requires an arid climate such as would be expected in the study area during an interglacial period. Because the pediment cuts old till deposits and is now above the current valley floor, we propose an Early Pleistocene interglacial age for the Pediment Gravels.

RIO TECKA VALLEY

1 TECKA DRIFT

A large morainic arc consisting of two or more adjacent terminal moraines is located at the eastern end of the upper Tecka valley. The deposit, which lies in the valley between Cerro Loma Alta and Chacras Pampa Chica, is 27 km long and about 5 km wide at altitudes of 800-950 m. The surface morphology is subdued, without an organized drainage system and the deposits are deeply weathered. Large erratic boulders are strewn over the surface. The moraine consists of large boulders (mostly granitic) reaching 2 m in diameter, cobbles and pebbles of granitoids and mesosilicic volcanics in a pale yellowish brown sandy-silty-clay matrix. Lateral moraines of the Tecka Drift lie on the side slopes of the Río Tecka valley at altitudes of 900-950 m, 25 m above the lateral moraines of the younger Mallin Grande Drift. The outwash of Tecka age is extremely modified by later fluvial processes and thus unrecognizable.

The Tecka Drift is located at similar altitudes and longitudes (71o W.) as the Pichileufu Drift recognized by Flint & Fidalgo (1968), 50 km to the north where the confluence of the Arroyo Pescado and the Río Tecka gives rise to the Río Gualjaina (see Figure 2). The Pichileufu glacial stage was considered as of the same age as the Illinoian? by Rabassa et al. (1987:262). However, the Tecka moraine loop is located immediately downstream of the well preserved terminal moraines of the last glaciation and therefore it must be considered equivalent to the El Condor glacial event (Flint & Fidalgo, 1968).

2 MALLIN GRANDE DRIFT

Two morainic arcs were recognized in the Río Tecka valley, which are considered to be the terminal moraine and a recessional moraine of the same glacial event. Both moraines have outwash terraces graded to them and glacilacustrine sediments behind the respective arc. The outer moraine, located at Pampa Grande, lies at an altitude of 750-800 m whereas the younger one, situated at Mallin Grande,

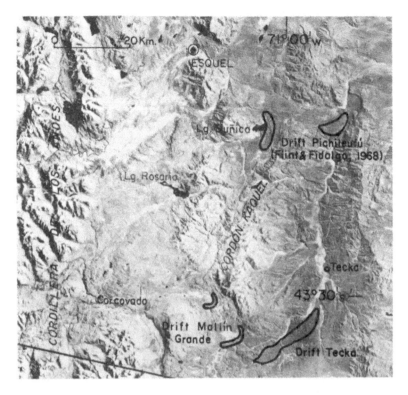

Figure 2. Glacial deposits on the Patagonian Cordillera at latitudes 43o00′and 43o30′S.

has an altitude of 800-850 m. The latter moraine is also the continental (watershed) divide. The morphology of both arcs is well preserved, except where the Río Tecka cuts through the older moraine. Some isolated boulders lie on the surface of the arcs. The till consists of granitic boulders which are striated in some cases, in a sandy-clay matrix of yellowish gray color. At a road-cut near Laguna El Sapo, overcompacted material has been identified as basal till. Lateral moraines are located at both sides of the valley on altitudes 800-875 m. Glaciofluvial stratified sands and gravels are located downvalley from the moraines. They can be followed 10 km downstream from the Pampa Grande arc and 2 km from the Mallín Grande arc. A flat surface underlaid by fine sediments is located behind each of the two arcs.

These sediments -fine sand, silt and clay- show few laminations and are considered glacilacustrine sediments, even though more observations are necessary to establish the nature of the lamination.

Mallin Grande Drift is found up valley from and topographically lower than the Tecka Drift. On the basis of the relative position in the valley and the well preserved moraine morphology, outwash and glacilacustrine deposits, the Mallin Grande Drift is correlated to the morainic loops at Lagos Panguipulli, Rinihue and Ranco (Illies, 1960 and Caviedes & Paskoff, 1975). On a regional basis, it correlates with the Nahuel Huapi glaciation (Flint & Fidalgo, 1968) which, on a worldwide scale is probably the equivalent of the Wisconsinan stage of North America (Rabassa, 1983).

ALTERNATING QUATERNARY CLIMATIC CHANGES

On the basis of the morphogenic systems described above and assuming a preservation of the entire record, an attempt can be made to establish some of the Quaternary climatic fluctuations of the Patagonian Andes at latitude 43o South.

Nothofagus remnants are found in sediments from the Patagonian Cordillera indicating a humid temperate-cold climate -similar to the current one- during the Late Tertiary. During Early Pleistocene times a very cold regime produced the onset of an ice sheet glaciation. As the climate warmed, the ice margins retreated and an arid to semi-arid climate followed which allowed the development of pediments on the Eastern front of the Andes. This semiarid period could be correlated with an interglacial stage. A new cooling lead to alpine glaciation during Mid Pleistocene times. Improvement of climatic conditions caused a glacier retreat, but a subsequent temperature lowering brought another ice advance in the valleys during the Late Pleistocene.

CONCLUSIONS

- A glacial stratigraphy and relative chronology is proposed for the Patagonian Andes at latitude 43o South, consisting of deposits of at least one older ice-sheet glacial stage and two distinctive younger alpine glacial episodes.

- An interglacial hanging pediment has been recognized on the eastern slope of the Patagonian Cordillera.
- Further work for precise absolute dating is necessary.

ACKNOWLEDGEMENTS

The authors wish to thank Dr. Jorge Rabassa, chairman of the organizing committee of the Final Meeting of IGCP 201 "Quaternary of South America", where part of this paper was presented. Field work has been partially supported by the Grant 5050145/83 of the Consejo Nacional de Investigaciones Cientificas y Técnicas (CONICET) of Argentina. Comments and observations of an anonimous reviewer were of considerable aid to highly improve the original manuscript.

REFERENCES

Caldenius, C.C. 1932. Las glaciaciones cuaternarias de la Patagonia y Tierra del Fuego. Dirección de Minas, Geología e Hidrología, Publicación 95. Buenos Aires.

Caviedes, C.N. & Paskoff, R. 1975. Quaternary Glaciations in the Andes of North-Central Chile. Journal of Glaciology,14(70):155-170.

Flint, R.F. & Fidalgo, F. 1964. Glacial geology of the east flank of the Argentine Andes between latitude 39 o10´S and latitude 41o20´S. Geol.Soc. Amer.Annl.Bull.335352.

Haller, M.J. 1981. Descripción geológica de la hoja 44 ab - "Trevelin". Servicio Geológico Nacional, unpubl. report. Buenos Aires.

Illies, H. 1960. Geologie der Gegend von Valdivia, Chile. Neues Jahrbuch fur Geologie und Palaontologie, 3(1):30-110.

Rabassa, J. 1983. INQUA Commission on lithology and genesis of Quaternary deposits: South America Regional Meeting, Argentina 1982. In Evenson, E.B., Ch. Schluchter & J. Rabassa (eds.) Tills and Related Deposits, 445-451. Rotterdam, A.À. Balkema Publishers.

Rabassa, J.; Evenson, E.; Schlieder, G.; Clinch, J.M.; Stephens, G. & Zeitler, P. 1987. Edad pre-Pleistoceno superior de la glaciación El Cóndor, Valle del Río Malleo-Neuquén,

República Argentina. Tenth Argentine Geological Congress, Proceedings, III:261-263. S.M. Tucumán.

Ramos, V.A. 1981. Descripción geológica de la hoja 47 ab - "Lago Fontana" Provincia del Chubut. Servicio Geológico Nacional, Boletín 183. Buenos Aires.

SIMONE SERVANT-VILDARY
Laboratoire de Géologie, Museum National d' Histoire Naturelle, Paris, France
KENITIRO SUGUIO
Instituto de Geociencias, University of São Paulo, São Paulo, Brazil

13

Marine diatom study and stratigraphy of Cenozoic sediments in the coastal plain between Morro da Juréia and Barra do Una, State of São Paulo, Brazil

ABSTRACT

Samples of sediments obtained from cores of four wells drilled by Nuclebrás, in the coastal plain of State of Sao Paulo, have been here studied from the viewpoint of diatom flora.

The irregular surface of the crystalline pre-Cambrian basement rocks is locally covered by the Pliocene Pariquera-Açu Formation - like deposits, whose contact with the Quaternary sediments is flat, being situated about 40 m below the present sea-level. The contact between the Holocene Santos Formation and the probable Pleistocene Cananéia Formation is very difficult to be recognized. It has been tentatively established based on two radiocarbon ages (Bah. 1138: 8,220 +/- 310 years B.P. and Bah. 1139: older than 30,000 years B.P., obtained from carbonaceous plant debris sampled respectively from wells F-003 and F-004). Dominantly clayey-silty intervals from the wells F-004 and F-006, probably related to the Cananéia Formation, have been selectively sampled and studied for definition of their diatom flora assemblage. According to these studies, there are littoral marine, estuarine and freshwater sediments, suggesting several phases of sea-level fluctuations. On the other hand, the most abundant species (**Raphoneis fatula**) is an extinct form and, until now, it had not been reported in sediments more recent than the Pliocene.

RESUMO

Amostras de sedimentos obtidas de testemunhos de quatro sondagens realizadas pela Nuclebrás, na planície costeira do Estado de São Paulo, foram aqui estudadas quanto ao seu conteúdo em diatomáceas.

A superfície irregular de rochas do embasamento cristalino pré-cambriano acha-se localmente coberta por sedimentos do tipo Formacao Pariquera Açu, de idade pliocênica, cujo contato com os sedimentos quaternários é plano, estando situado cerca de 40 m abaixo do nível do mar atual. O contato entre a Formação Santos, de idade holocênica, e a provável Formação Cananéia, de idade pleistocênica, é muito difícil de ser reconhecido. Ele foi tentativamente estabelecido com base em duas datações ao radiocarbono (Bah. 1138 : 8.220 +/- 310 anos A.P. e Bah. 1139: $>$ 30.000 anos A.P., obtidas de restos vegetais carbonizados amostrados respectivamente dos poços F-003 e F-004).

Intervalos predominantemente síltico-argilosos dos poços F-003 e F-004, provavelmente relacionados a Formação Cananéia, foram seletivamente amostrados e estudados para determinação da assembléia de diatomáceas. De acordo com esses estudos, existem sedimentos marinhos litorâneos, estuarinos e de água doce, sugerindo várias fases de flutuações do nível marinho. Por outro lado, a espécie mais abundante (Raphoneis fatula) é uma forma extinta e, até agora, não foi registrada em sedimentos mais novos do que o Plioceno.

INTRODUCTION

The sedimentary deposits here studied occur at the northeastern extremity of the Cananéia-Iguape coastal plain (Figure 1) defined by Suguío and Martin (1978).

During the climax of the Cananéia transgression, about 120,000 years B.P., the sea reached the foot of the Serra do Mar coastal ranges. This episode is characterized by shore-face clayey-sandy deposits followed by foreshore sands, both covering the continental Pariquera Açu Formation, which has been assumed as Pliocene in age (Sundaram & Suguio, 1983). The retreat of the sea was accompanied by the deposition of beach

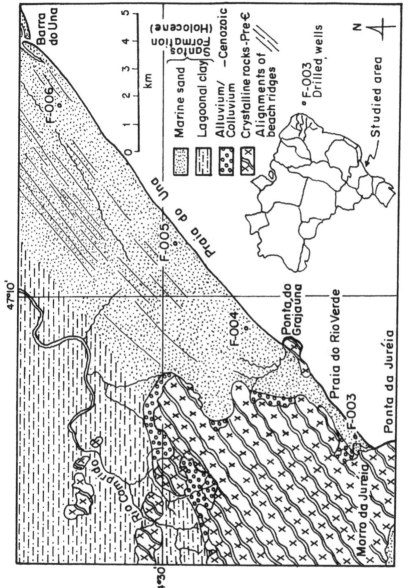

Figure 1. Geologic map showing the locations of drilling sites.

ridges overlying the transgressive sands. During
the northern hemisphere Würm glacial stade, the
sea-level dropped until more than 100 m below the
present level, when the Cananéia Formation surface
was strongly dissected by a drainage net. Until
this point, the evolutionary history of this
coastal plain is quite similar to that of its
southern portion.

The Santos transgressive sea deeply encroached
the more inland portions between Morro da Juréia
(Juréia hill) and Barra do Una (Una outlet). This
fact is demonstrated by the occurrence of several
shell-middens, dominantly composed of **Anomalocardia
brasiliana**, which normally lives within muddy bay-
bottom sediments (Suguío and Martin, 1978). A
barrier-island, probably developed soon after the
5,100 years B.P. maximum of the Santos
transgression, separated the bay from the open-sea,
which has been transformed successively into a
lagoonal area, freshwater coastal lake and finally
to a swampy lowland. A shell-midden situated at the
Rio das Pedras headwaters, with 3,800 years B.P.
radiocarbon age (Suguío & Martin, 1978) is
indicating the presence of the sea at least until
that time. Then, the communication of this lagoonal
area with the open-sea has been interrupted only
after the second Holocene maximum sea-level, which
occurred about 3,400 years B.P.

The four drilling wells, here studied, belong
to a Nuclebrás geological investigation program
performed for the geotechnical characterization of
the future site for thermonuclear power plants
(Hassano et. al., 1984).

DRILLED LITHOSTRATIGRAPHIC UNITS

Six rotary drilling wells, with a disposition
parallel to the present shoreline, were perforated
by the Nuclebrás in the studied area, whose depths
changed between 80 m and more than 150 m. Only the
drilling wells F-003, F-004, F-005 and F-006
(Figure 1 and 2), with more clayey-sandy beds and
organic plant debris, have been sampled,
respectively, for diatom studies and radiocarbon
datings.

The local pre-cambrian crystalline basement is
represented by the Brasiliana Age Complexo Costeiro
(Coastal Complex), composed of migmatitic-granitic
rocks grading to schists, which is characterized by

a very irregular paleo-relief, as demonstrated by the wells that reach the substrate. Meanwhile the substrate is about 50 m deep in some places, like at the sites of the wells F-003 and F-004, the well F-005, distant only 4,1 km from F-004, does not reach the substrate until its final depth of 150 m. These depressions could be explained as paleo-valleys excavated by the drainage net, during the periods of lower sea-levels, as well as by possible faults trending parallel to those of Rio Verde area (Morro da Juréia), oriented according to N 32 oW to 87 oW directions, and probably related to the Guapiara alignment (Hassano et al., 1984).

The crystalline rocks are overlain by continental deposits of the Pariquera Açu Formation, probably Pliocene in age, whose thickness is very changeable according to the substrate's paleo-topography. Its basal portion is characterized by coarse sands and gravels grading upward to reddish-coloured conglomeratic layers alternated with sandy beds and clayey-silty matrix.

Apparently, the top of the Pariquera Açu Formation is flat and is situated about 40 m below the present sea-level, exhibiting a very sharp contact with the superimposed sedimentary deposits. It is very difficult to distinguish the Pleistocene Cananéia Formation from the Holocene Santos Formation, but certainly both are present in the area, as suggested by radiocarbon ages obtained from carbonized plant debris sampled from the wells F-003 (Bah. 1138 : 8,220 +/- 310 years B.P.) and F-004 (Bah. 1139 ➤ 30,000 years B.P.). The first age is clearly indicative of Holocene age and the second one is related to a minimum age. The limit between the Pleistocene and the Holocene formations has been tentatively established in the studied drilling wells (Figure 2), which indicated a thickness variable between 14 and 27 m for the Cananéia Formation, and between 13 and 30 m for the Santos Formation. Inlandwards in the lagoonal area, northwestward from the area of the Figure 1, the top of the Cananéia Formation mapped by Suguío and Martin (1978) is situated about 10 m above the present sea-level, where its thickness could be more than 50 m.

As in the type-locality of Cananéia, the homonymous formation in this area is more clayey-sandy in its basal part, becoming more sandy (fine to very fine sands) on the top. The Santos

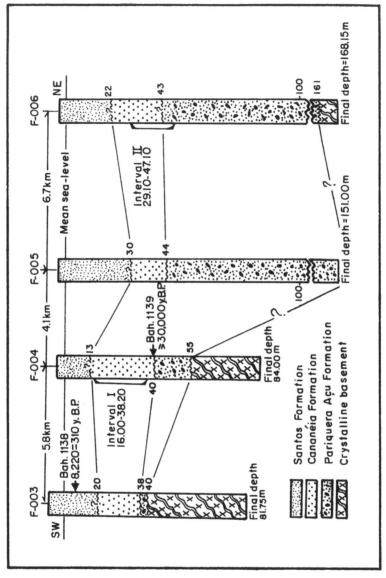

Figure 2. Columnar sections of four wells drilled by Nuclebrás with indications of radiocarbon ages and studied intervals.

Formation comprises, besides shallow marine sands, fluvial, lacustrine and mangrove deposits.

DIATOM BIOSTRATIGRAPHY

Samples for diatom study have been taken from two cores of the wells F-004 and F-006 (Figure 2). In the first one (F-004), seven samples have been analyzed, but only three of them contain a very rich diatom flora. In the second one (F-006), also seven samples have been analyzed, however only three of them contain a poor diatom flora. Assemblages of both cores are similar; a detailed systematic study has been done only in the richest samples F-004/4 and F-004/7 (Table 1).

1 VARIATIONS OF THE DIATOM ASSEMBLAGES ALONG THE CORES FROM THE WELL F-004

The sample 4/1 from about 12 m below the present sea-level in the Holocene Santos Formation does not contain any diatom; this lack of diatoms is perhaps due to the sandy nature of the sediment.

Between 13 and 40 m, six samples have been taken from the Pleistocene Cananéia Formation. Only three of these samples contain diatom; samples 4/4 and 4/7 show a true littoral marine assemblage whereas sample 4/6 contains a freshwater to slightly saline water diatom flora.

The sample 4/2 from about 16 m below the present sea-level is the uppermost sample of the Cananéia Formation. It contains some broken frustules of **Synedra ulna** and **Melosira arenaria**. Both are freshwater species, probably related to a lower sea-level period.

The sample 4/3 from 20.80 m of depth does not contain diatoms.

The sample 4/4, from 24.15 m of depth, is the most interesting sample due to the good preservation of the frustules and to the greatest specific diversity. The diatom assemblage is characterized by true marine species, **Raphoneis fatula** (37%) (Figure 3) and **Paralia sulcata** (32%) (Figure 4) being dominant.

The sample 4/5, from 31.45 m of depth, does not also contain any diatom.

The sample 4/6, from 34.65 m of depth, showed the diatom flora distributed within a very thin (1

Table 1. Frequency of the diatom species in some samples from the wells F-004 and F-006, with ecological information.

Species \ Samples	4/2	4/4	4/6	4/7	6/2	6/3	6/5	Ecology
Actinocyclus tenellus	0	1	0	0	0	0	0	M.W.
Actinoptychus kusnetzkianus	0	0	0	1	0	0	0	M.*
Actinoptychus senarius	0	4	0	0	1	1	0	M.C.
Actinoptychus splendens	0	2	0	2	1	1	0	M.W.
Actinoptychus undulatus var. tamanicus	0	1	0	2	0	0	0	M.*
Actinoptychus vulgaris	0	0	0	2	0	0	0	M.E.
Biddulphia reticulata	0	2	0	0	0	0	0	M.W.
Cerataulus smithii	0	1	0	1	0	0	0	M.W.
Coscinodiscus concavus var. minor	0	0	0	1	0	0	0	M.*
Coscinodiscus divisus	0	0	0	4	0	0	0	M.
Coscinodiscus perforatus	0	2	0	0	0	0	0	M.C.
Delphineis surirella	0	0	0	1	0	0	0	M.C.
Diploneis bombus	0	2	0	1	1	0	0	M.
Endictya japonica	0	1	0	0	0	0	0	M.*
Melosira arenaria	2	0	0	0	0	0	0	F.
Navicula cincta	0	0	4	0	0	0	0	B.
Navicula directa	0	0	2	0	0	0	0	M.
Nitzschia granulata	0	0	0	1	2	0	0	M.B.
Nitzschia tryblionella var. victoriae	0	0	0	1	1	0	0	M.B.
Paralia sulcata	0	4	0	1	2	2	1	M.B.
Pinnularia microstauron	0	0	4	0	0	0	0	F.
Podosira stelligera	0	3	0	0	0	1	0	M.W.
Raphoneis fatula	0	4	0	1	2	0	0	M.*
Stauroneis amphioxys	0	0	1	0	0	0	0	M.B.
Synedra ulna	1	0	0	0	0	0	0	F.
Thalassiosira eccentrica	0	0	0	3	0	0	0	M.
Thalassiosira nodulolineata	0	2	0	3	0	0	0	M.
Xanthiopyxis sp.	0	0	0	2	0	0	0	M.*

M.= Marine, B.= Brackish, F.= Freshwater, W.= Warm, C.= Cold, E.= Eurythermal, 0 = Absent, 1 = Very rare, 2 = Rare, 3 = Abundant, 4 = Very abundant and * = Only reported as fossil.

mm thick) yellow silty lamina overlying a very hard and black sediment without diatoms. This sample contains a lacustrine slightly saline water diatom assemblage. The dominant species is **Pinnularia microstauron** (Fig. 6: 45), which is oligahalobious. **Navicula cincta** (Fig. 6: 46) oligo-mesohalobious and **Stauroneis amphioxys** (Fig. 6: 49) mesohalobious species are rare. There are also few marine species, e.g., **Navicula directa**. This assemblage indicates a lagoonal freshwater or slightly saline water environment, probably related to a short period of sea-level drop.

The sample 4/7, from 38.20 m of depth, contains a rich and well-preserved diatom flora. All the species are marine, most of them are composed of still living species, except for **Coscinodiscus concavus** (Fig. 4: 24) and

274

Xanthiopyxis. The Pennatophycideae are represented by the following marine species: **Nitzschia tryblionella var. victoriae** (Fig. 6: 50), **Diploneis bombus** (Fig. 5: 37), **Nitzschia granulata** (Fig. 3: 7), **Delphineis surirella** (Fig. 6: 48), all these species widespread in ancient deposits and still living today. According to Andrews (1981 a, b) Delphineis surirella indicates shallow marine brackish waters along coasts of cool to temperate climate.

2 VARIATIONS OF THE DIATOM ASSEMBLAGES ALONG THE CORES FROM THE WELL F-006

When compared with the cores of the well F-004, the frequence of the frustules as well as its diversity are lower. Seven samples have been taken from 21.80 to 130.50 m of depth.

The sample 6/1, from a depth of 21.80 m, does not contain any diatom. The sample 6/2, from a depth of 29.10 m, contains some **Paralia sulcata**, **Actinoptychus splendens**, **Raphoneis fatula** and **Nitzschia granulata**. The sample 6/3, from a depth of 33 m, contains some **Podosira stelligera**, **Paralia sulcata** and **Actinoptychus sp.** The sample 6/4, from a depth of 43.40 m, does not contain any diatom. The sample 6/5, from a depth of 77.10 m, contains some **Paralia sulcata**. The samples 6/6 (130.50 m) and 6/7 (130.50 m) do not contain any diatom.

DIATOM TAXONOMY AND GEOLOGY

1. Actinocyclus tenellus (Breb.) Andrews
    Description - Diameter = 26 μm. The Brazilian form is quite similar to that one illustrated by Abbott and Andrews (1979) in their plate 1:9, except by the presence of six sectors instead of five. The single radial rows delimiting the sectors are composed of big areolae (8 in 10 μm), the areolae filling the sectors are smaller (10 in 10 μm), arranged in radial irregular rows. The margin is occupied by very small areolae. Marginal labiate processes are located at the end of the single radial rows of bigger areolae.
    Geological range - Miocene to Recent.
    Ecology - Navarro (1982:12) found this species in warm waters, and Hendey (1964:84) in the neritic zone.

2. **Actinoptychus kusnetzkianus** Pantocsek (Fig. 5: 36)

Description - Diameter = 30μm. The surface of circular valve is divided into six sectors. When compared with A. senarius, the valve is flatter, the differences between raised and depressed sectors are slight, the pores are smaller, and the reticulate network is less apparent. Three of six setors are bordered by a hyaline area near the margin as it occurs in A. **vulgaris**. A labiate process is located at the middle of the margin of each sector. The Brazilian form looks quite similar to Pantocsedk´s drawing in Tafel XXVI, Figure 383, band III. Another species, A. **punctulatus** illustrated in Schmidt´s Atlas, Tafel 109, seems to be a close species.

Geological range - Never has been reported after Pantocsek (1886), who found it in marine Kusnetzk deposits of Tertiary age from USSR.

3. **Actinoptychus senarius** (Ehrenberg) Ehrenberg (Fig. 5: 29-35).

Description - Diameter = 25 to 40 μm. It is divided into six sectors. In internal view, the three depressed sectors bear a labiate process near the margin, the pores are fine and arranged in irregular, linear rows, the three elevated sectors do not bear labiate processes, the pores are bigger, and arranged in irregular rows (Fig. 5: 32-34). The mantle contains seven rows of small pores arranged in a quincuncial pattern (Fig. 5: 35). In external view, a reticulate network clearly appears under LM (Fig. 5: 29). The external openings of the labiate processes are short, broad tubes, located near the margin of the valve (Fig. 5: 31).

Geological range - Cretaceous to Recent. Common in Neogene deposits.

Ecology - It is common in cool waters of modern seas, being able to tolerate considerable changes of salinity (G.W. Andrews, personal communication). It is a north-temperate species, frequently found in the plankton of deep waters (Wornardt, 1967:44).

3. **Actinoptychus splendens** (Shadbolt) Ralfs (Fig. 6: 39-41).

Description - Two different morphological types can be distinguished, separated by their size and some different features. The first one groups the big specimens (diameter = 70 μm) with eighteen alternating raised and depressed sectors, which fit the type species (Fig. 6: 39). The second one groups the small specimens (diameter = 30 µm), characterized by only twelve sectors, a stellate central hyaline area (Fig. 6: 41), and a coarse reticulate network, but the strongest difference appears in the elevation of the raised sectors. In the first type, the elevation is the same along the sector from the margin to the center. In the second type, the elevation is highest near the margin, this restricted part of the raised sectors appears out of focus in valve view (Fig. 6: 40, arrow), the labiate process is located in this zone. This same peculiar feature appears on **Actinoptychus aequalis** (Andrews & Abbot, 1985), plate 11, and on **Polymyxus coronalis** - L.W. Bailey (Syn. Actinoptychus) from the Pará river in the slide no490 of the Tempere and Peragallo Collection, illustrated by Schmidt, Tafel 132.

Perhaps it will be necessary to separate within the **Actinoptychus** genus, species with regular raised sectors and species with irregular raised sectors, and not to attribute these small **Actinoptychus** to species **splendens**, as for the big ones.

Geological range - Late Miocene to Recent (Andrews, 1980).

Ecology - According to Hendey´s opinions (1964:95), it is a common littoral form, frequent in the English channel. For Navarro (1982), it is a sublittoral form of warm waters, and John (1983) found it in the summer within planktonic and benthic samples.

5. Actinoptychus undulatus var. tamanicus, Jousé (Fig. 6: 42-44).

Description - Length = 60 µm and width = 45 µm. It is an oval diatom with two very elevated sectors (Fig. 6: 43), and two slightly elevated sectors (Fig. 6: 44), bearing a labiate process each one (see schematic drawing), separated by narrow depressed sectors. The pores are round and arranged in regular parallel rows.

A form similar to that of Tempere and Peragallo collection (2nd edition, n°
182) was found in the Middle-Late Miocene Richmond Formation of Virginia. However, in the plate n o 18 of the Ehrengerg Mikrogeologie (1838), concerning to this formation, this form is not illustrated, and it is missing too in Andrews´ recent work on the same formation published in 1986. This form is illustrated only in "the diatoms of USSR" edited by Glezer et al.

(1979), under the name A. undulatus var. tamanicus
(Plate 40:13).

Probably, Actinoptychus undulatus var.
tamanicus can be considered as synonymous of
Polymyxus coronalis, because of the areolae
arrangement on the valve face and the restricted
elevation near the margin of the raised sectors,
although it has a lower number of sectors.

Geological range - Miocene of the Richmond
Formation, according to Tempere and Peragallo.
Marine Miocene deposit of Kuriles Islands,
according to Glezer et al. (1979:178), where it is
associated with true Miocene assemblage
characterized by Stenphanopyxis inermis,
Thalassiosira zabelinae, Denticula kamtschatica,
but also with Paleogene species, like
Pseudopodosira hyalina, Coscinodiscus payeri var.
payeri and Hemiaulus elegans.
6. Actinoptychus vulgaris A. Schmidt (Fig. 6: 38).

Description - It is a very peculiar species
with only eight sectors and a characteristic
pattern. There is no reticulate network, pores are
big in regular linear rows, with the same
disposition and number in the depressed and
elevated sectors. In external view, the elevated
sectors are delimited by a hyaline area near the
margin. The closest form was illustrated by John
(1983, Plate X, Figure 10). A nice specimen was
found in Tempere and Peragallo Collection (2nd
edition, no 622, from the Malacca
detroit, China).

Ecology - It is a benthic and planktonic form,
pH around 8, and temperature between 13
o and 35oC (John, 1983).
7. Biddulphia reticulata, Ropper (Fig. 3: 6).

Description - Length = 45 μm and width = 30
um. Five hexagonal areolae in the center and eight
at the extremities. There are no peculiar remarks
about the morphological features of this species.

Ecology - It presently lives in sublittoral
marine environments, predominantly in warm waters
(Navarro, 1982:14).
8. Cerataulus smithii, Ralfs (Fig. 3: 8-9).

Description - Diameter = 30 μm. Circular valve
with two small triangular elevations, round areolae
inside, with external velum consisting of volae, 8
in 10 μm (Fig. 3: 9).

Ecology - According to Navarro (1982:17), it
is a sublittoral to supralittoral species, common
throughout the year. For Hendey (1964:106) it is

278

common and widespread littoral species along all North Sea coasts. Also widespread in the eastern Mediterranean sea during the Early Holocene, where was replaced by **Paralia sulcata** during the Late Holocene (Sneh et al., 1986).

9. Coscinodiscus concavus var. minor, Grove (Fig. 4: 24).

Description - Diameter = 30 μm. Circular valve, not flat. Focus cannot be done on the entire valve at the same time. In one half part of the valve the areolae (7 in 10 um) appear round bordered by a thick rim; this image corresponds to the internal morphology of the areolae, meanwhile, on the other half, the areolae appear hexagonal, each angle marked by a siliceous round point clearly seen under LM (outside morphology of the areolae).

Similar forms have been found in the slides n o 569 and 570 of the Tempère and Peragallo Collection from Miocene Otago Oamaru Formation of New Zealand; quite identical to the brazilian specimen.

Geological range - Until now, only reported in Miocene deposits.

10. Coscinodiscus divisus, Grunow (Fig. 4: 25,26).

Description - Diameter = 52 μm. Circular valve containing seven polygonal areolae in 10 μm, with external velum visible under LM. Valve fasciculate, divided into twelve sectors. This form is morphologically very similar to the specimen illustrated by Abbott and Andrews (1979, Plate 2, Figure 3, p.236), and was identified as C. divisus by Andrews (written communication).

Geological range and ecology - According to these authors: "previously reported from rocks of Pliocene age to Holocene in a widespread modern littoral and neritic environments".

11. Coscinodiscus perforatus, Ehrenberg (Fig. 3: 10).

Description - Diameter = 46 μm. Circular valve with rounded areolae, bigger in the center of the valve (5 in 10 μm) than near the margin. The marginal strutted processes are regularly spaced every 2 μm (Fig. 3: 10, arrows) there are five central strutted processes irregularly spaced, located at about midway between the center and the margin.

Geological range - Miocene to Recent.

Ecology - It is frequent in the plankton of the North Sea (Hendey, 1964:77).

12. Endictya japonica, Kanaya (Fig. 4: 23).
    Description - Diameter = 18 μm. It is a small circular species. The areolae are loculate, and covered by an hexagonal cellulation (5 in 10 μm in the center), smaller near the margin of the valve.
    Geological range - It has been described by Kanaya (1959) in the Middle Miocene of deposit Onnagawa Formation in Japan and by Hendey (1981) in the uppermost part of core 10, site 278, leg 29, in the Subantarctic Southwest Pacific.
13. **Paralia sulcata** (Ehrenberg) Cleve (Fig. 4: 11-19).
    Description - It constitutes about 32% of the assemblage observed in the sample no
4/4 of the well F-004. Its greatest interest is due to the morphological features. Widely studied by Crawford (1979), SEM observations on Brazilian material supplied with additional information, which are very useful to better understand the valve structure of this complicated species. A detailed description will be given at the end of this chapter.
    Geological range - Throughout the Tertiary to Recent. Widespread in Miocene deposits; it is common in the Richmond Formation, Virginia (Ehrenberg, 1838, Mikrogeologie, Plate 18). This formation has been recently studied by Andrews (1986, Plate 7:12-14).
    Ecology - This species is very important for paleoecological reconstructions. According to Andrews, this species is common in modern shallow marine coastal waters; to Hendey (1964:73) and
John (1983:16) in shallow marine environments and to Navarro (1982) in sublittoral zone.
14. **Podosira stelligera** (Bailey) Mann (Fig. 4: 20,21).
    Description - Diameter = 15 to 50 μm. Strongly convex circular valve, whose central portion presents sparse areolae, marginal area fasciculate with radiate fine rows of areolae.
    Ecology - For Hendey (1964:90), it is a tychopelagic species; for Andrews (1980:32), it is a common planktonic form in Recent marine and coastal waters. Navarro (1982:11) found it in sublittoral zone with a temperature of 17 oC.
15. Raphoneis fatula, Lohman (Fig. 3: 1 to 5).
    Description - Length = 40 to 60 μm and width = 8 to 10 μm. Flat, lanceolate, more or less rounded valve, with transversal rows of puncta (7 in 10

μm), longitudinal rows of puncta (7 - 8 in 10 μm). Very thin axial area. Specimens very changeable in shape and morphology, which have been firstly related to several species, but according to Andrews, these morphological variations cannot be considered as specific differences, thus they have been reported as a single species R. fatula, Lohman.

Geological range - R. fatula has been described by Lohman (1938) from the Pliocene San Joaquin Formation, east flank of North Dome, Kettleman Hills, Kings Country, California. It has not yet been reported in modern environments. Very abundant in the well F-004, its presence in this level makes questionable the problem of attribution to the Pleistocene of the Cananéia Formation. Alternatively, the sediments assumed to be of Cananéia Formation could be older than this formation, because 14C dating supplied only with its minimum age.

Ecology - The habitat of the genus is shallow marine environment.

16. **Thalassiosira eccentrica** (Ehrenberg) Cleve (Fig. 4: 27,28).

Description - Diameter = 35 to 45 μm. Circular valve with rounded areolae inside and hexagonal outside, arranged in tangential rows with large sectors not very clearly marked in some specimens (8 in 10 um). Many strutted processes are scattered throughout the valve, a strong labiate process can be observed under LM (Fig. 4: 27), marginal strutted processes appear in Figure 4: 28.

Geological range - Lower Miocene to Recent.

Ecology - This species is found in neritic plankton (Andrews & Abbott, 1985). It is a common planktonic diatom found in large numbers at the lower stations during the summer in Australia (John, 1983:19).

17. **Thalassiosira nodulolineata** (Hendey) Hasle & Fryxell (Fig. 4: 22).

Description - Diameter = 25 μm. Circular valve with hexagonal loculate areolae (6 in 10 μm) in linear rows. Marginal processes can be observed under LM, but they are not visible on Fig. 4: 22. Six small central strutted processes are arranged in a small sub-circular rim.

Geological range - Unknown to Recent.

Ecology - Marine.

18. Genus **Xanthiopyxis**
This genus only appears in the sample n°
   4/7 of the well F-004. There are few
specimens, three different dominant forms should
possibly related to X. lacera, Forti (Late Miocene
to Early Pliocene), X. diaphana, Forti (Middle
Miocene to Early Pliocene), X. new sp., Wornardt
(Early Pliocene).

MORPHOLOGICAL REMARKS ON THE SPECIES **PARALIA
SULCATA**

Paralia sulcata is a very interesting species from
different points of view: complexity in its
structure, abundance in modern environments and its
great geographical-geological distribution in
ancient sediments. It has been illustrated by
previous authors under different names, for
example, by Schmidt (Tafel 175, 176, 177). The last
more detailed study on this species is due to
Crawford (1979).
      In the Brazilian material this species is
abundant. Detailed SEM observations allowed us to
clarify some yet problematic aspects concerning the
ultrastructure of the valve face, essentially the
relation existing between the different images
appearing under LM and the location in the chain of
the different images of the valve face.
      Crawford (1979) has shown that the
structures formerly attributed to **Paralia sulcata**
and to P. **sulcata var.** **coronata** represent the same
species, since they occur together in the same
chain. It is a case of heteromorphy, named by this
author as **Paralia sulcata** structure, Type 2, and
Paralia sulcata var. coronata, Type 1.
      In the Brazilian material, both structures can
be observed in big specimens (diameter = 50 μm), as
well as in small ones (diameter = 12μm). Type 2 is
more complex than type 1. Under Lm three different
morphologies (or images) can be distinguished and
described; they depend on the location within the
chain of the observed valve, and the side through
which they are charged. Figure 7 illustrates the
following different possibilities:
      a) Sub-type 2A (Crawford, 1979: Figure 6, 17).
From the margin to the center it is possible to
observe: 1) the spines, 2) a rim of big pores, 3) a
rim of stellate radiating markings, and 4) a
central circular hyaline area. This image

282

corresponds to the external view of the linking cells (Figure 7: 4).

b)Sub-type 2B (Crawford, 1979. Figure 9) (Fig. 4: 13, 15, this paper). Under LM, it is composed of: 1) the spines, 2) a rim of big pores, 3) a rim of fine striae at the same level, 4) a rim of stellate radiating markings and 5) a central circular hyaline area. This image corresponds to the internal view of the linking cell (Figure 7). Under SEM, this internal view appears very simple (Fig. 4: 11-13) and according to Crawford (1979: Figure 28) it is composed of: 1) a ring of fine striae near the margin of the valve, the rimoportulae are located near the mantle edge, and 2) a hyaline circular central area.

The stellate radiating markings and the rim of big pores are not visible because they are strouded by an internal lamina (Fig. 4: 13) composed, from the center to the margin, of a hyaline lamina covering the radiating markings, and a rim of fine striae covering the big pores. When by chance this internal lamina is broken (Fig. 4: 15 and Crawford, 1979: Figure 20), it is possible to see the big pores and the radiating markings which are below it.

So, the two different images (sub-types 2A and 2B) observed under LM, represent the same kind of valve face, regarded in internal or external view respectively (Figure 7:4 and 3).

c) Sub-type 2C. The third image of the valve face (Fig. 4:19) is quite difficult to interpret. It is composed of: 1) the spines, 2) the rim of big pores, 3) the stellate radiating markings, 4) the rim of fine striae, and 5) a rim of short wedge-shaped markings.

The presence at the same time of the radiating markings of the linking cells and the wedge-shaped markings of the separation cell can only be explained, if we admit that we observe the internal view of the last linking valve, located just before the separation valve (Figure 7:2). The wedge-shaped markings appear by transparency under LM. This case is not so common, that is normal because the number the last linking cell linked with the separation cell, is smaller compared with the number of the linking cells in the chains.

EXPLANATION OF THE PLATES

Figure 3: 1-10

LM: Scale bar = 10 μm

1 - 5: **Raphoneis fatula**, Lohman
      1 - Same specimen under different focus.
      2 - Dyssymmetrical form
      3 - Quadrate pattern of ornamentation
      4 - Interior of valve
      5 - Detail of the same specimen, showing
         the external valve.
   6: **Biddulphia reticula**, Ropper
   7: **Nitzschia granulata**
8 - 9: **Cerataulus smithii**, Ralfs
      8 - LM at different magnifications.
      9 - SEM view of the internal valve.
  10: **Coscinodiscus perforatus**, Ehrenberg
      Internal valve at different magnifications.
      Arrowhead:
      Marginal strutted process. Arrow: Central
      strutted process.

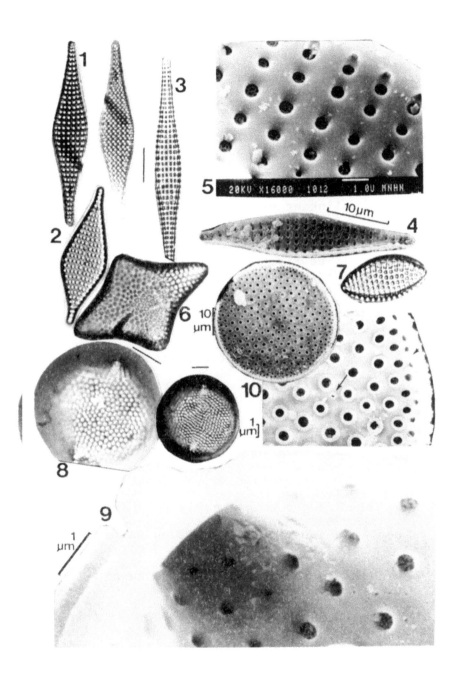

Figure 3: 1-10

Figure 4: 11-28

LM: Scale bar = 10 um

11 - 19: **Paralia sulcata** (Ehrenberg) Cleve
   11 - SEM: Lateral and internal views of one valve of linking cell.
   12 - LM: Lateral view of a chain of three linking cells.
   13 - SEM: Internal view of a valve of linking cell or of the hypovalve of a separation cell.
   14 - LM: Internal view of a linking cell. Image corresponding to sub-type 2B (See Fig. 7:3).
   15 - SEM: Internal view of a valve of linking cell, similar to 13, but with an internal broken lamina (arrowhead), which allows to see the radiating markings and the big pores.
   16 - LM: Chain of **Paralia** including a complete separation cell (arrow), a complete last linking cell (double arrows) and the hypovalve of another linking cell.
   17 - LM: Internal view of the epivalve of separation cell. (See Fig. 7:1b - Type 1).
   18 - SEM: External view of the epivalve of a separation valve (See Fig. 7:1a - Type 1).
   19 - Internal view of the epivalve of the last linking cell, including the internal view of the epivalve of the separation valve (See Fig. 7:2 - Type 2C).
20 - 21: **Podosira stelligera** (Bailey) Mann
   Two specimens of different sizes.
   22: **Thalassiosira nodulolineata** (Hendey) Hasle & Fryxell.
   23: **Endictya japonica**, Kanaya.
   24: **Coscinodiscus concavus var. minor**, Grove.
25 - 26: **Coscinodiscus divisus**, Grunow.
27 - 28: **Thalassiosira eccentrica**, (Ehrenberg) Cleve.

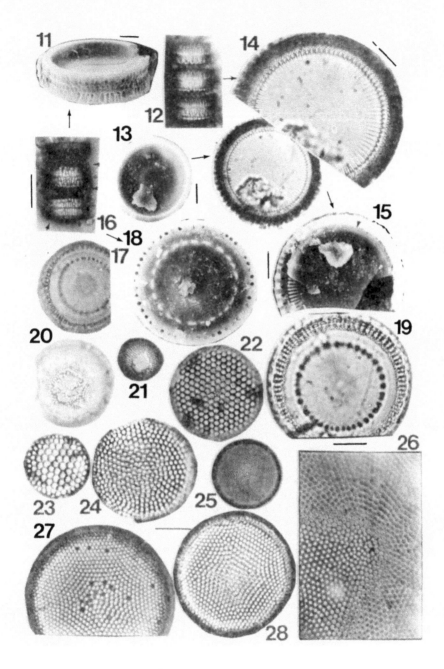

Figure 4: 11-28

Figure 5: 29-37

LM: Scale bar = 10 um

29 - 35: **Actinoptychus** **senarius** (Ehrenberg)
Ehrenberg.
29 - LM: (a) Focus on raised sectors, (b)
Focus on depressed sectors.
30 - SEM: External valve.
31 - Detail on the external reticulate
system and the external foramen of
labiate process.
32 - SEM: Internal valve.
33 - SEM: Detail on depressed sector, with
irregular rows of pores on a raised
sector, bearing a labiate process.
34 - SEM: Detail of the labiate process.
35 - SEM: Rows of small pores on the
mantle arranged in quincuncial pattern.
36: **Actinoptychus kusnetzkianus**, Pantocsek.
37: **Diploneis bombus**, (Ehrenberg) Cleve.

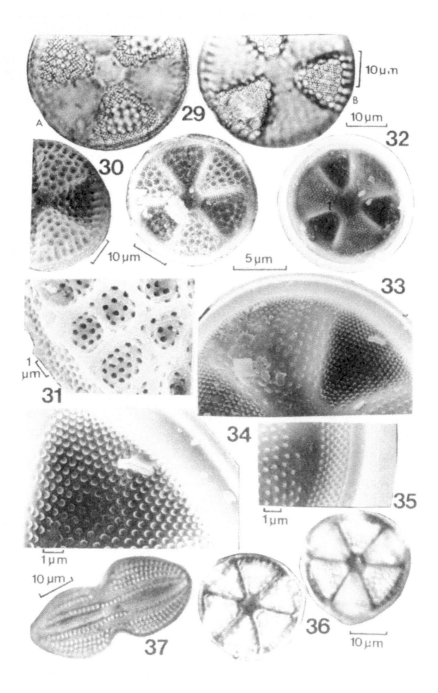

Figure 5: 29-37

Figure 6: 38-50

LM: Scale bar = 10 um

      38: **Actinoptychus vulgaris**, A. Schmidt
         a = Focus on raised sectors, b = Focus on
         depressed sectors.
      39: **Actinoptychus splendens**, (Shadbolt) Ralfs
40 - 41: **Actinoptychus splendens**, (Shadbolt) Ralfs
         40 - Stellate central area (See localized
         great elevation near the margin in
         the raised sectors, appearing out of focus
         and bearing the labiate process).
         41 - Detail of the central area.
42 - 44: **Actinoptychus undulatus** var. **tamanicus**,
         Jousé.
         42 - Entire valve.
         43 - Detail of raised terminal sectors.
         44 - Detail of intercalated raised
            sectors.
      45: **Pinnularia**   **microstauron**,   (Ehrenberg)
         Cleve.
      46: **Navicula cincta**, (Ehrenberg) Kutzing
      47: **Nitzschia punctata**, (W. Smith) Grunow
      48: **Delphineis surirella**, (Ehrenberg) Andrews
      49: **Stauroneis amphioxys**, Gregory
      50: **Nitzschia** **tryblionella** var. **victoriae**,
         Grunow

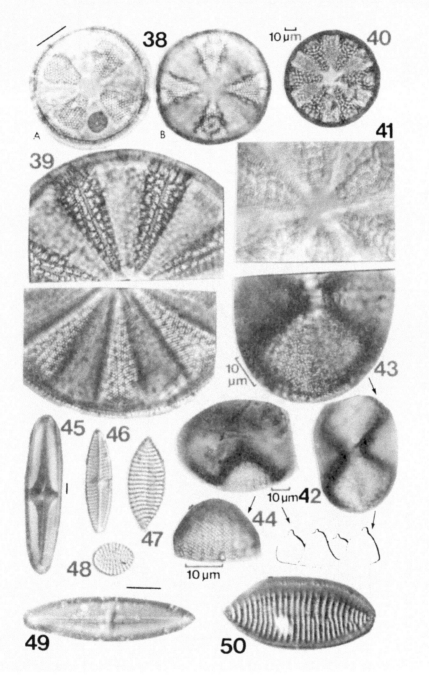

Figure 6: 38-50

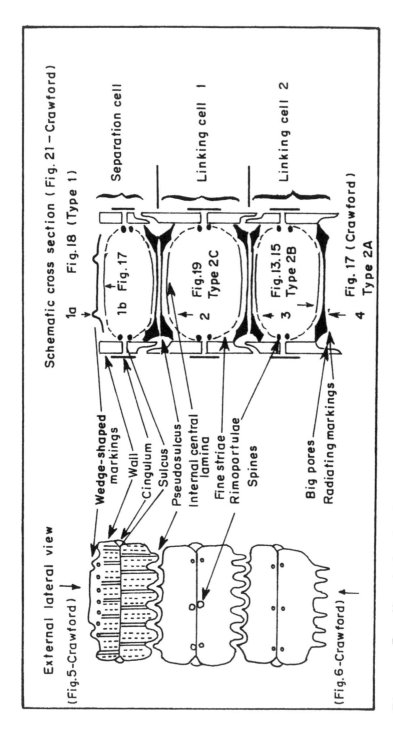

Figure 7: Morphological remarks on the species *Paralia sulcata*

d) Type 1 (Fig. 4: 18). It corresponds to the
view of the epivalve of the separation cells,
formerly called **Paralia sulcata** var. **coronata**. It
is composed of: 1) a rim of big pores and 2) a rim
of wedge-shaped markings (If it is regarded from
the outside - Figure 7: 1a and Crawford, 1979:
Figure 15:5. If it is regarded from the inside -
Figure 7:1b, the big pores are recovered by a rim
of fine striae corresponding to the marginal
structure of the internal lamina). The internal
view of the hypovalve of the separation cells
presents the same image as the valves of the
linking cells and can be assigned to sub-type 2B.

In summary, the variability in the structure
of the valve face of **P. sulcata**, which induced the
former authors to create many different varieties,
can be understood taking into account the internal
and external views of the two different types of
cells (separation and linking cells) and the
different types of valves (epi and hypovalve of the
separation cells) in the same chain.

FINAL CONSIDERATIONS

Four wells drilled in the coastal plain, between
Morro da Juréia and Barra do Una (State of São
Paulo, Brazil), reached the Pre-cambrian
crystalline basement, cutting before the
continental Pliocene Pariquera Açu Formation, and
shallow marine Cananéia and Santos formations, both
of Quaternary age. The Cananéia Formation,
according to the contained diatom flora, was
deposited within a littoral marine environment with
interbedded level of lacustrine diatoms indicating
phases of sea-level drop. The analyzed samples from
the Holocene Santos Formation do not contain
diatom.

The Cananéia Formation exhibited a rich and
well preserved diatom flora, most of the species
constituted by living forms, but six of them are
extinct, and the most abundant one (**Raphoneis
fatula**) has not been reported until now in
sediments more recent than Pliocene. The presence
of these extinct species raises the problem of the
age of the Cananéia Formation. Alternatively, the
sediments assumed to be of this formation could
belong to older deposits.

ACKNOWLEDGEMENTS

Thanks are due to Dr. G. Andrews for his constructive suggestions on specific determinations, and to Dr. A. Ehrlich who read critically the manuscript. The authors are also indebted to the geologist João Hilário Javaroni (General Superintendent of the Mineral Research and Prospection Department), as well as to the geologists Sumio Hassano and Samir Saad, all of them from Nuclebrás (Brazil), who supplied with samples and furnished helpful informations.

REFERENCES

Abbott, W.H. & Andrews, G.W. 1979. Middle Miocene marine diatoms from Hawthorn Formation within the Ridgeland Trough, Carolina and Georgia. **Micropaleontology**, 25(3):225-271.

Andrews, G.W. 1979. Miocene marine diatoms from the Choptak Formation, Calvert County, Maryland. **U.S. Geol. Survey Prof. Paper** 910:26p.

Andrews, G.W. 1980. Neogene diatoms from Petersburg, Virginia. **Micropaleontology**, 26(1):17-48.

Andrews, G.W. 1981a. Revision of the diatom genus **Delphineis** surirella (Ehrenberg). G.W. Andrews, n. comb., **6th Diatom Symposium** Proc., Ed. Ross. Otto Koeltz. Science Publishers: 81-90.

Andrews, G.W. 1981b. Revision of the diatom genus **Delphineis** and morphology of **Delphineis** surirella (Ehrenberg). G.W. Andrews, n. comb., **6th Diatom Symposium** Proc., Ed. Ross. Otto Koeltz. Science Publishers: 81-92.

Andrews, G.W. 1986. Miocene diatoms from Richmond, Virginia. **Journal of Paleontology**, 60(2):497-538.

Andrews, G.W. & Abbott, W.H. 1985. Miocene diatoms from the Formation, Thomas County, Georgia, **Bull. American Paleontology**, 87(321):109p.

Crawford, R.M. 1979. Taxonomy and frustules structure of the marine centric diatom **Paralia sulcata**. J. Phycol., 15:200-210.

Cupp, E.E. 1943. Marine plankton diatoms of the west coast of North America. Univ. of California Press, Berkeley and Los Angeles, **Reprint O. Koeltz Science Publishers**, 236p.

Foucault, A.; Servant-Vildary, S.; Niaqiao, F. & Powichrowski, L. 1986. Un des plus anciens gisements de diatomées découvert dans l'Albien-Cenomanien inférieur des Alpes Ligures (Italie). C.R. Acad. Sc. Paris, 303, II, 5:397-402.

Glezer, S.I. et al. 1979. The diatoms of USSR. Otto Koeltz Science Publishers. Reprint, 1:400p.

Hasle, G.R. & Fryxell, G.A. 1977. The genus Thalassiosira species with a linear areola array. 4th Diatom Symposium Proc., Ed. Simonsen J. Crames, 54:15-66.

Hassano, S.; Muller, M. & Stein, J.H. 1984. Estudos geológicos na regiao do Rio Una (Município de Iguape, SP) para implantacao de centrais nucleares. XXXIII Congr. Bras. Geol. Anais 1:2089-2100, Rio de Janeiro.

Hendey, N.I. 1964. An introductory account of the smaller algae of British coastal waters. Her Majesty's Stationary Office, Part V: Bacillariophyceae, 317p.

Hendey, N.I. 1981. Miocene diatoms from the subantarctic Southwest Pacific. Deep Sea Drilling Project Leg 29, Site 278, Core 10, Bacillaria, 4:65-124.

John, J. 1983. The diatom flora of the Swan river estuary, Western Australia. Biblioteca Phycologica, J. Cramer., 64:359p.

Kanaya, T. 1959. Miocene diatom assemblages from the Onnagawa Formation and their distribution in the correlative formations in northeast Japan. Sci. Rep. Tohoku Univ., 2nd Series (Geology), 30:1-130.

Lohman, K.E. 1938. Pliocene diatoms from the Kettleman Hills, California. U.S. Geol. Survey Prof. Paper, 189-C:81-102.

Makarova, I.V. 1981. Principles of the systematics of Thalassiosira, Cleve and the significance of its taxonomic characters. 6th Diatom Symposium Proc., Ed. Ross. Otto Koeltz Science Publishers, 1-14.

Navarro, J.N. 1982. Marine diatoms associated with mangrove prop. roots in the Indian river, Florida, USA. Biblioteca Phycologica, J. Cramer, 61:151p.

Pantocsek, J. 1886. Beitrage zur Kenntnis der fossilen Bacillarien ungarns, Berlin, NW 5 Ed. W. Junk, 1903, Teile 3, Marine Bacillarien, 77p.

Rodrigues, L. 1984. Contribuicao ao conhecimento das diatomáceas do Rio Tubarao, Santa Catarina, Brasil. Insula, Florianópolis, 14:47-120.

Sneh, A.; Weissbrod, T.; Ehlrich, A.; Horowitz, A.; Moshkovitz, S. & Rosenfeld, A. 1986. Holocene evolution of the northeastern corner of the Nile Delta. Quaternary Research, 26:194-206.

Souza Mosmann, R.M. de. 1984. Levantamento preliminar das diatomáceas (Chrysophyta, Bacillariophyceae) na regiao de Anhatomirim, Santa Catarina, Brasil. Insula, Florianópolis, 14:2-46.

Suguio, K. & Martin, L. 1978. Quaternary marine formations of the State of Sao Paulo and southern Rio de Janeiro. Special Publ. no1, 1978 International Symposium on Coastal Evolution in the Quaternary, 55p., Sao Paulo.

Sundaram, D. & Suguio, K. 1983. Nota preliminar sobre uma assembléia miofloritica da Formacao Pariquera-Acu, Estado de Sao Paulo. VIII Congr. Bras. Paleontologia, 1983, MME/DNPM, Série Geologia N o 27, Paleont./Estrat. No 2:503-505.

Tempere, J. & Peragallo, H. 1915. Diatomées du monde entier. Collection Tempere & Peragallo, 2eme édition, 480p.

Van Landingham, S.L. 1985. Potential Neogene diagnostic diatoms from the western. Snake River Basin, Idaho and Oregon. Micropaleontology, 31(2):167-174.

Wornardt, W.W. 1967. Miocene and Pliocene marine diatoms from California. Ocassional Papers, California Academy of Sciences, 63:108p.

MARTIN H.IRIONDO
CONICET, Paraná, Argentina

14

Map of the South American Plains – Its present state

ABSTRACT

The main objective of the Map of the South American
Plains Project is the elaboration of a map in scale
1:5,000,000. A general characterization of the
major sedimentary systems of the plains is also
made. The first phase of the project was dedicated
to fit a definition and a classification of plains
and to solve several methodological problems. Later
on, more than a half of the plains of the continent
have been already mapped. The map will be finished
in 1990.
          Several basic properties of plains are
outlined: a) they are composed of a few large
sedimentary systems; b) the dynamical processes are
relatively slow; c) the sedimentary deposits have a
large lateral extension and relatively small
thickness; d) there are frequently great climatic
variations in a single plain; e) the sedimentary
processes produce homogeneous mineralogy and
granulometry in large areas. Two important
sedimentary systems are sketched: the Pilcomayo
alluvial fan and the Pampean aeolian system. The
Pilcomayo fan was built during a sequence of humid
and dry climates, whereas the Pampean system is of
a periglacial nature.

RESUMEN

El Proyecto Mapa de las Llanuras Sudamericanas
tiene como objetivos la elaboración de un mapa en
escala 1:5.000.000 y la caracterización general de
sus grandes sistemas. Se halla en un avanzado

estado de desarrollo. Más de la mitad de las llanuras del continente fueron mapeadas, después de elaborarse una definición y una clasificación. Se espera terminar en 1990. Las grandes llanuras poseen ciertas propiedades básicas: a) están formadas por unos pocos sistemas sedimentarios de grandes dimensiones; b) las variaciones climáticas suelen ser considerables dentro de una misma llanura; c) la mineralogía y granulometría de los sedimentos suele ser homogénea en grandes extensiones. Se describen dos importantes sistemas de llanura: el abanico aluvial del Pilcomayo y el sistema eólico pampeano. El abanico del Pilcomayo fue formado en una sucesión de climas húmedos y secos; el sistema pampeano tiene origen periglacial.

INTRODUCTION

The Map of the South American Plains Project, a part of the IGCP-201, tries to fill a considerable lack of knowledge about large regions of our continent. Plains cover millions of square kilometers in South America. In spite of their great theoretical and economical value, such areas are improperly known. Some of them, as the Llanos of Bolivia and Ecuador, are covered by the less explored forests of the world. Others, like the Argentine Pampa, have been exploited and populated only since last century, and studied by Quaternary scientists in only a few aspects.

The little knowledge available about plains appears to be generated by two negative factors: a) the lack of a clear definition and of an operative classification of plains; b) the frequent division of the large plains in two or more administrative territories (countries, provinces), which produces a fragmentation of the existing surveys.

The main objective of the present project is the elaboration of a wall map (Wandkarte) of the South American plains in scale 1:5,000,000. It should serve as a cartographic basis for more detailed studies. The project has also led us to the preliminary characterization of the large morphogenetic and sedimentary systems of the plains.

METHODOLOGY

The applied methodology is based on the interpretation of Landsat images in scale 1:1,000,000, with the support of aerial photographs and large-scale satellite imagery in critical areas. As a complement, field work and a general characterization of granulometry and mineralogy of sediments are made in the mapped units. Studies made by other authors resulted of great value in some regions, specially those published by Radambrasil in Brazil and the geomorphologic maps of several Argentine provinces, published by different Government agencies.

DEFINITION AND CLASSIFICATION

A definition and a classification of plains were fitted during the first phase of the project (Iriondo, 1986). A plain is defined as an area of Earth surface with small or no relief, where the local topographic elements are more significative than the regional slope. Water, particularly, shows a characteristic behaviour: the runoff is relatively low in comparison with evaporation and infiltration; the hydrographic systems show low efficiency for carrying water.

The sedimentary processes of plains tend to create local relief: sand dunes, fluvial levees, moraines, etc. The absolute elevation of such landforms is normally modest, but in those extremely horizontal regions they have a first-order significance.

According to the above definition, a plain can be placed at any elevation above the sea level. The nature of the borders of the plains are not considered; a plain can end at a mountain, in the sea margin, at a cliff or against any other geomorphological element.

The classification is genetical, hierarchical and open. Its first level separates the areas with a general tendency to sinking (Aggradation Plains) from those areas with a general trend to elevation (Structural Plains). Aggradation plains are clearly dominant in South America. The second and the third level of the classification related to aggradation plains are shown in Figure 1.

Considering the scale of the map and the size of the sedimentary systems found in the South

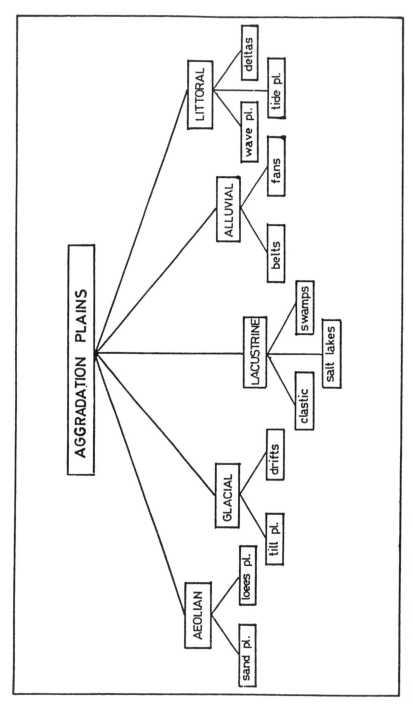

Figure 1. Classification related to aggradation plains

American plains, the third level of the classification was used for the cartographic representation (loess plains, deltas, etc.).

PRESENT STATE OF THE PROJECT

This project was started in 1983. It was supported almost exclusively by CONICET (the National Research Council of Argentina) from the beginning. At present it is in an advanced state of development. The first phase of the work consisted in solving several theoretical and methodological problems, such as definition, classification or scales of interpretation.

The second phase comprised the mapping of Patagonia, Pampa, Chaco and smaller plains of Argentina and Paraguay. It was finished in 1987. The map of the Argentine (1:4,000,000) and Paraguayan (1:1,000,000) plains were published in restricted editions.

The third phase is being performed. It comprises the mapping of the alluvial fans located *at the eastern slope of the Andes. They are very large units, reaching the western regions of Brazil in some cases. The dense and homogeneous forest partially masks the geomorphological features, posing some particular methodological problems.

The final phase will be dedicated to the mapping of the Brazilian plains and plateaus, the Atlantic coastal plain and the plains linked to the Guyana shield. The project will be finished in 1990.

RESULTS

I - BASIC CHARACTERISTICS OF LARGE PLAINS

More than a half of the plains of South America have been already mapped or examined to some degree in our study, covering from the equator to a latitude of 55o S. The observed patterns, complemented with information obtained elsewhere, allow the identification of some basic characteristics of large plains:
a) A large plain is formed by a few sedimentary systems.

b) The sedimentary systems are simple; they have very large lateral extension and reduced thickness. Examples of such peculiar pattern are the fluvial Ituzaingó Formation in NE Argentina (120,000 sq.km in extension and 10 to 20 m of thickness) and the Upper Pleistocene loess of the Pampa (300,000 sq.km in extension and 5 to 10 m thick).

c) The processes of transport and sedimentation produce a high degree of mixing of the minerals of the source rocks and granulometric selection, resulting in homogeneous granulometry and mineralogy of large areas.

d) There are frequent climatic variations in a single plain. The Chaco is semiarid in the west (600 mm/yr) and humid in the east (1200 mm/yr); the Pampa has a tropical climate in the north (19o C mean annual temperature) and a temperate climate in the south (15o C m.a.t.).

e) The physical geography (geology, climate, geomorphology) of the surrounding regions is dominant in many plains, especially those controlled by hydric processes. The 80% of the sediments of the Amazon plain comes from Andean and Sub-andean regions; the Paraná river at Rosario is in equilibrium, not with the neighboring areas, but with the geologic and climatic characteristics of Sao Paulo (Brazil), a region located 2500 km to the northeast.

f) The dynamic processes in plains are slow. River flow, migration of dunes and analogous phenomena are considerably slower than in other comparative environments.

II - TWO SIGNIFICATIVE SEDIMENTARY SYSTEMS

Several major sedimentary and morphological systems were identified and, to some degree, described during the development of the project. The most probably important ones are the Pampean aeolian system and the Pilcomayo alluvial fan.

1 THE PILCOMAYO ALLUVIAL FAN

The Pilcomayo river forms the largest alluvial fan in Chaco (Figure 2). Besides, it is the unique major active fan in South America. The apex is located at Villa Montes (Bolivia), at the foot of the Aguarague mountain range; its distal line runs along the Paraguay river, 750 km to the east. The total surface of the fan is 210,000 sq.km covering part of Bolivia, Argentina and Paraguay.

The age of the fan is unknown, but it can be assumed not younger than Upper Tertiary; then the Pilcomayo is antecedent to the Aguarague range. The fan is formed by a few main sedimentary units in surface. In the West, a formation composed of quartz silt and very fine quartz sand is present; the grains are coated by hematite and large plates of interstitial illite. The sediment is yellowish brown in colour, compact and moderately plastic. This unit is characterized by numerous ephemeral channels, 5 to 15 km of visible length and irregular pattern. It reaches approximately longitude 60o W.

To the east, a unit composed of 10 to 20 m thick swamp clays, extends to the Paraguay river. An irregular N-S fracture appears at the boundaries between both units. The surface of the eastern unit is covered by permanent and semi-permanent swamps. It is crossed by W-E alluvial belts, 3 to 15 km wide, formed by large paleochannels and lateral levees. Inside the paleochannels the present-day rivers of the Chaco Oriental flow, clearly smaller in size, forming an underfit pattern (Cucchi, 1973).

During the last centuries the Pilcomayo river flowed into the large swamp Patiño, at the Argentina-Paraguay border, 250 km west of the Paraguay river. The mechanism of colmatation was studied by Cordini (1947), who registred high concentrations of suspended load (up to 40,000 ppm) and the formation of dikes of trunks and other vegetal debris during the floods.

Several years ago the depression became completely filled up. Since 1980, the river is undergoing a slow process of avulsion, filling up the lower portion of the channel with very fine sand and silt and spilling out the water to the surrounding plain. Such a mechanism causes receding of the spill-out section upstreams in successive floods. The rate of receding is 10 to 35 km per year. The colmatation mechanism is as following: Discharge varies from 3600 cu.m/sec in floods to 80 cu.m/sec in floods to 80 cu.m/sec in low waters (Rabicaluc, 1986). The normal channel is 500 m wide upstream of the colmatation area; in the lower reach narrow ditches cut the levees (10 ditches were found in 45 km), draining each of them a part of the water discharge.

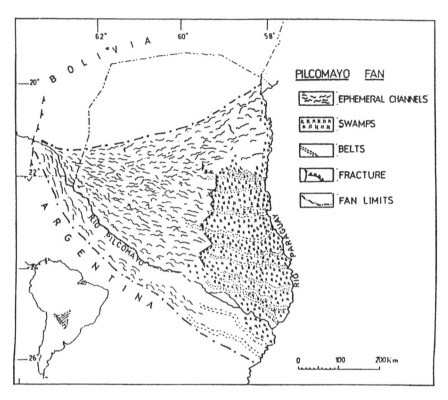

Figure 2. The Pilcomayo alluvial fan.

Most of the sediments remain in the channel.
Consequently, the river narrows steadily downstream
until it completely disappears, leaving a filled-up
channel, at the same level than the surrounding
plain. The water spilled to Argentina formed a
large swamp, 250 km long and 7 to 12 km wide. Its
dynamics and morphology are paludal. The waters
derived to Paraguay produced a similar phenomenon.
    In a first approach, the building of the
Pilcomayo fan has been a sequence of two contrasted
mechanisms: a) the development of stable fluvial
belts during humid periods, like the present
climate; b) a widespread sedimentation through a
system of ephemeral channels in dry periods - in
both cases, the development of swamp in sunken
blocks. Interpreting the map of Figure 2, the
oldest surficial unit is the eastern swamp clays-
and-fluvial-belts, followed by the western

ephemeral-channels-formation, which covers it
partially. The present fluvial belt (humid period)
cuts the younger formation (dry climate).

2 THE PAMPEAN AEOLIAN SYSTEM

The Pampean aeolian system covers more than
600,000 sq.km in central Argentina (Figure 3). The
southern part of the system is formed by a large
sand sea, 300,000 sq.km in area, from Buenos Aires
to Mendoza. The thickness of the sediment is small,
between 5 and 15 m. Most of the sand is fine to
very fine, yellow to yellowish brown in colour.
Potassium felspar (about 50% of the total),
volcanic glass and quartz are the dominant minerals
in southern Córdoba (Sánchez and Blarasin, 1987).
Several fields of longitudinal dunes can be
observed, although dissipation processes masked the
original forms. Some dunes are more than 200 km
long. One of the largest dune fields is located in
northwestern Buenos Aires; the dunes of such area
are particularly long and have a gentle anti-clock
curvature.
The limit of the sand sea follows an irregular
SSE-NNW direction through the provinces of Buenos
Aires, Santa Fé and Córdoba. Behind a transition
belt, 5 to 15 km wide, a loess formation covers the
rest of the Pampa and neighbouring areas. It is
composed of friable silt, yellowish brown in
colour, with abundant concretions of calcium
carbonate. Plagioclases and lithic fragments
predominate in the associated sand fraction in
southern Santa Fé (INTA, 1984). Illite is the
dominant clay mineral; the ratio quartz/felspars is
less than 1 in the silt fraction. The original
thickness of the formation (5 to 10 m) is preserved
in most of the region; on the other hand, the
outcrops in the river cliffs show only a 2 to 3 m
thick loess.
To the northeast (northern Santa Fé and Chaco)
the typical loess passes transitionally to a fringe
of loessoid deposits, sedimented in swamp
environments. Such sediments are brown to greenish
gray in colour, with calcium carbonate concretions
and manganese minerals in concretions and pores
(Iriondo, 1987). Pleistocene megamammals are
frequently found in the loess formation and in the
sand sea.
The Pampean system was formed during the last

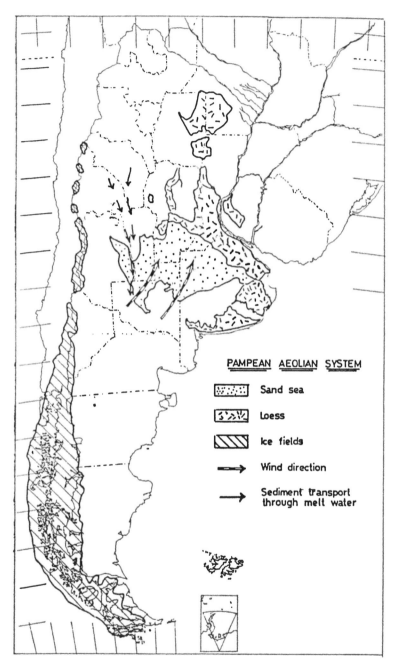

PAMPEAN AEOLIAN SYSTEM

Sand sea
Loess
Ice fields
Wind direction
Sediment transport through melt water

Figure 3. The Pampean eolian system.

glacial maximum of the Andes. The ice field covered a large area south of 28o S. (Clapperton, 1983), allowing the occurrence of an anti-cyclon, which produced SSW-NNE winds. The rest of the Argentine Cordillera, north of 28o S., was almost free of ice. The reason of that was the severe dryness of the climate, rather than a moderate temperature. Such an environment becomes very efficient in producing silt, fine sand and illite through physical weathering. The sediment was transported to the south along the Cordilleran piedmont by the Bermejo-Desaguadero-Atuel fluvial system (Figure 3), and blowed up by the SSW wind in the lower Desaguadero (southern Mendoza and La Pampa provinces).

In that region, near the northern border of Patagonia, the sand sea began to develop in cold desert conditions. The fields of longitudinal dunes make evident the desertic climate and the abundant supply of sand. The anti-clock deviation of the dunes in Buenos Aires is coherent with the anti-cyclonic circulation of winds in the Southern Hemisphere.

To the NNE, crossing the Río Salado of Buenos Aires, the climate in the plain was peridesertic, allowing the precipitation and fixing of the dust transported in suspension by the wind, and forming the loess mantle. Approximately 300 km more towards the NE, along the Río Salado of Santa Fé, the climate changed to subhumid; the plain was characterized by extensive swamps. Then, the aeolian dust deposited in such an environment.

REFERENCES

Clapperton, Ch. 1983. The glaciation of the Andes. Quaternary Sciences Review, 2:83-155.
Cordini, R. 1947. Los ríos Pilcomayo en la región del Patiño. Dirección de Minas y Geología, Anales I, 83pp.
Cucchi, R. 1973. Los ríos desajustados de Formosa. Rev. Asoc. Geológica Argentina. 28:156-164.
INTA, 1984. Carta de Suelos de la República Argentina. Hoja 3363- 30 Berabevú. INTA, 141pp, Buenos Aires.
Iriondo, M. 1987. Geomorfología y Cuaternario de la provincia de Santa Fé. D´Orbignyana, 3 (in press), Buenos Aires.

Rebicaluc, H. 1986. Situación del río Pilcomayo. Dirección de Recursos Hídricos de Formosa, Internal Report, 5pp.

Sánchez, M & Blarasin, M. 1987. Depósitos eólicos cuaternarios de la zona de Cuatros Vientos, Depto Río Cuarto, Prov. de Córdoba. **X Congreso Geológico Argentino**, Actas III: 293-296.

MÓNICA C.SALEMME
Vertebrate Palaeontology Division, La Plata Museum, CONICET, La Plata, Argentina

15

Zooarchaeological studies in the humid Pampas, Argentina

ABSTRACT

This paper studies the adaptative strategies of the Pampean hunter-gatherers during the Holocene. It is a part of a larger multidisciplinary project, which involves not only archaeologists but also geologists, palaeontologists and palinologists.

Faunal remains coming from archaeological sites located in different environments of the Buenos Aires province, were analyzed from two points of view: 1) palaeozoogeographical-palaeoenvironmental and 2) cultural. The total of the remains was associated with lithic and ceramic materials and sometimes with human burials. The study was based on the taxonomic assignment of the bones, the MNI estimation, analysis of anthropic modification on bones and reviewing of the present corology of the species registered in the sites. Hypotheses about the palaeoenvironments in which the hunters had to be adapted since the Early Holocene until historical times were formulated.

RESUMEN

Este trabajo estudia las estrategias adaptativas de los cazadores- recolectores pampeanos durante el Holoceno. Esta contribución es parte de un proyecto multidisciplinario mayor, el cual involucra no sólo arqueólogos, sinq también geólogos, paleontólogos y palinólogos.

Restos faunísticos provenientes de sitios arqueológicos ubicados en distintos ambientes de la Provincia de Buenos Aires han sido analizados desde dos puntos de vista: 1) paleozoogeográfico-paleoambiental y 2) cultural. La totalidad de los restos fueron hallados asociados con materiales liticos y cerámicos, y a veces, con enterratorios humanos. El estudio fué basado en la asignación taxonómica de los huesos, la estimación del NMI, el análisis de la modificación antrópica de huesos y la revisión de la corología actual de las especies registradas en los sitios. Se han formulado hipótesis acerca de los paleoambientes a los cuales los cazadores tuvieron que adaptarse desde el Holoceno temprano hasta los tiempos históricos.

INTRODUCTION

The aim of this paper is to summarize the information related to zooarchaeological studies in the Pampean Region (Buenos Aires province sector) from two points of view. First, the palaeoenvironmental aspects, which could have affected the corology of some faunal species and second, the aboriginal adaptation to those palaeoenvironmental conditions and the strategies of subsistence employed.

Basically, the lifeway in the Pampas from the Late Pleistocene until the XVII century was based on hunting, and complementary, on gathering of various resources. Therefore, it is interesting to know in which way the adaptive strategies were changing during such a long period according to the availability of certain species.

In fact, this subject will be analized taking into account not only the archaeological information, but also chronicles of the voyagers that travelled across the present extent of the Buenos Aires province during the XVI, XVII, and XVIII centuries (ethnohistorical information), besides geological, palaeontological and palinological data.

Environmental aspects

The Pampean Region is a large plain extending from the foothills of the Andean cordillera to the west to the Atlantic Ocean to the east and from the Sierras Centrales and Río Paraná in the north to the Río Colorado in the south. It comprises about 600.000 sq.km that may be divided into "Humid

310

Pampa" and "Dry Pampa" (Daus, 1946). In this paper, archaeological sites studied are located in the Humid Pampa or Proper Pampa, which embraces the main part of the Buenos Aires province. This sector is separated from the Dry Pampa by the 600 mm isohyet.

The climate of the Humid Pampa is temperate with a warm period between November and March. Rainfall decreases from the east to the west (1.000 to 600 mm mean annual precipitation). Only two mountain ranges interrupt the pampean plains: the Ventania and Tandilia systems, with maximum elevation of 1247 m a.s.l. and 550 m a.s.l., respectively.

Phytogeographically, Pampean plains belong to the "Provincia Pampeana" (Cabrera, 1971), characterized by a gramineous steppe or pseudosteppe, without native autochtonous trees. However, the Spaniards described some native thorny species like "caldén" and "algarrobo", besides "sauce criollo" (red willow) along the river banks upon their arrival in the XVIth Century. From a zoogeographical point of view, the Humid Pampa corresponds to the Pampean Dominion (Ringuelet, 1955; 1961), in which mammals and birds are characteristic of open environments, although there is a great influence from the Subtropical Dominion.

Several subareas can be distinguished within the Humid Pampa (Fig.1), with different characteristics for each of them, not only from the environmental aspects but also from the archaeological viewpoint (see Salemme, 1987; Politis y Salemme, i.p.). In this sense, it is clear that the faunal remains received scarce or none attention since the beginnings of the archaeological Pampean studies. Except F. Ameghino (1880), who emphazised the findings of faunal remains in sites along the Río Luján (in the northeastern of Buenos Aires province), the other researchers did not pay attention to the bone remains. It was approximately ten years ago when the interest on the faunal wastes arised and then the study of prehispanic economy, techniques of hunting, palaeoenvironments, etc., began emphatically.

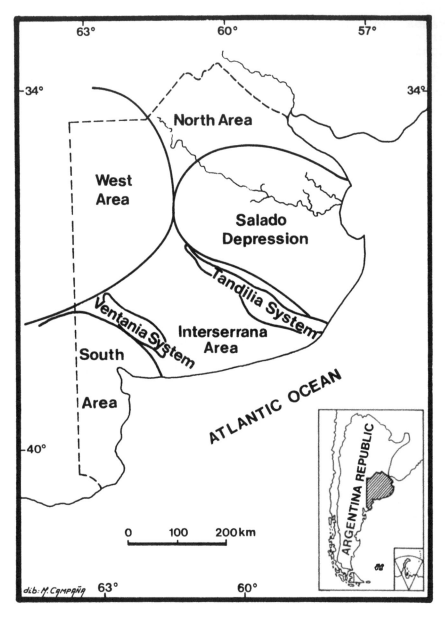

Figure 1. Geographical subareas in the Humid Pampas

ARCHAEOLOGICAL DATA

The information summarized in this paper comes from
two kind of analytical levels: 1) faunal remains
obtained from collections and/or our own
excavations; 2) faunal information processed and
published by other researchers, who have excavated
their sites using similar methodologies. Therefore,
and as a complement, the faunal comments included
in the chronicles of the XVI to XIX centuries are
also considered.

The main point taken into account was the
location of the sites ; they are localized in
different subareas, which show distinct
environmental characteristics at present. The goal
was to find out how the environment was in those
subareas in the past and consequently how the
aboriginal behaviour was in such
palaeoenvironments. The sites under my own analysis
were 1) Cañada de Rocha, 2) Río Luján Site (both of
them located in the Ondulate Pampa), 3) Tres Reyes
1, in the Interserrana area and 4) La Toma, close
to the Ventania hilly area. The information coming
from other authors corresponds to the following
sites: 5) Lechiguanas Island (Delta sector), 6)
Cañada Honda (Ondulate Pampa), 7) La Moderna
(northeastern of the Interserrana area), 8) Fortín
Necochea (northwest of the Interserrana area); 9)
Tixi Cave, 10) Cerro La China, 11) Lobería 1 (all
of them located in the Tandilia hilly area); 12)
Zanjón Seco Locality, 13) Cortaderas and 14) Arroyo
Seco S2 (the three of them in the southeast of the
Interserrana area) (Fig.2).

The faunal remains were studied analyzing the
previous bibliography, if it existed, and comparing
them with reference collections. The study was
based on the taxonomic classification of the bones,
the counting of the minimun number of individuals
(MNI), the analysis of the remains with
anthropogenic modifications (bone implements) and
the analysis of present distribution of the species
registered in the sites.

A brief summary of each site is necessary to
know the faunal and cultural contexts and their
chronologies (for more details see Salemme, 1987
and the bibliography cited there). The location of
the sites are shown in Fig. 2 and the species
recovered in every site are shown in Tables 1 and
2, which are separated in the sites under my own
analysis and the sites studied by the authors, as

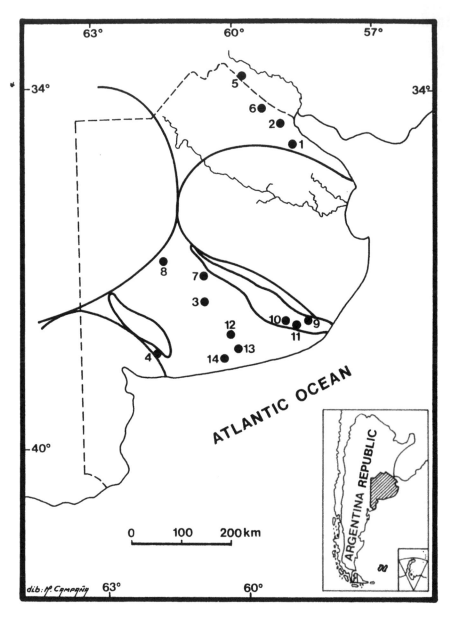

Figure 2. Location map of the sites.

314

it was explained before. Likewise, the chronological position of the sites is shown in Tables 3 and 4.

1) Cañada de Rocha is located in the Ondulate Pampa, close to a creek bank which goes downstream to the Luján river. The site was excavated by F. Ameghino (1880); he recovered a great amount of pottery sherds and faunal remains but a very few lithic implements. He described the geology and the stratigraphic position of the remains and emphasized on the species which were used by the aborigines; most of them were not living in the area at the end of the XIXth century (Salemme, 1983) (see list of species in Table 1).

2) Río Luján Site is an open-air site located along the Río Luján 90 km farther north than Cañada de Rocha, also in the Ondulate Pampa (see Fig.2). Two different sectors were excavated, one of them a place of human burials, and the other a campsite, where a few lithic artifacts and a great amount of pottery and faunal remains were recovered (Petrocelli, 1975). Except a piece assigned to Lama sp., the bones belong to a several species characteristic of the Subtropical Dominion (Salemme and Tonni, 1983; Salemme, 1987) (see Table 1).

3) Tres Reyes 1 is a multicomponent open-air site, close to a small lake in the Interserrana area. The human occupations took place during prehispanic times and several cultural differences among the Components were detected (Politis and Madrid, 1988; Madrid and Salemme, 1988) but they could not be observed in the faunal remains. In fact, the same species are registered through the entire sequence; so, it has been proposed that the guanaco (main resource) and the Pampean deer (complementary resource) were the species incorporated into the basic diet during the Middle and Late Holocene in this site (Salemme, 1987) (see Table 1).

4) La Toma is an open-air site localized in the Sauce Grande river banks, in the surroundings of the Ventania hilly area. The Central and Patagonian Dominions have a big influence on that area because of their proximity, though it is possible to find some species of Brazilian center like "mulita" (Dasypus hybridus) or "cuis" (Cavia aperea). The indigenous occupations detected took place under different environmental conditions; whereas the earliest one seems to have inhabited under an arid or semiarid cycle in the Middle Holocene, the latest human occupation (995 yr B.P., Table 3)

315

Table 1. Species recovered in sites 1, 2, 3 and 4.-

Taxon	C.de Rocha	S.R.Luján	L.T.Reyes	La Toma CS	NI
Adelomedon brasiliensis	-	-	-	1	-
Zidona dufresnei	-	-	-	1	-
Volutidae	-	-	-	-	-
Doradidae indet.	-	x	-	-	-
Siluridae indet.	x	-	-	-	-
Colossoma mitrei	-	1	-	-	-
Bufo cf. paracnemis	-	-	1	-	-
Ceratophrys sp.	-	-	1	-	-
Chelyidae indet.	-	1	-	-	-
Tupinambis cf. teguixin	x	6	-	-	1
Rhea americana	-	1	-	2	-
Rheidae	x	-	1	-	-
Ceryle torquata	-	1	-	-	-
Fulica leucoptera	-	1	1	-	-
Gallinula sp.	-	1	-	-	-
Chauna torquata	-	1	-	-	-
Ajaia ajaja	-	1	-	-	-
Columbinae	x	-	..	-	-
Nothura sp.	x	-	..	-	-
Ardea cocoi	x	-	-	-	-
Vanellus chilensis	x	-	-	-	-
Phoenicopterus	x	-	-	-	-
Coscoroba coscoroba	x	-	-	-	-
Cavia aperea	x	6	-	1	-
Ctenomys sp.	x	3	4	4	3
Dolichotis patagonum	x	-	-	2	1
Hydrochoerus hydrochaeris	-	-	1	-	-
Myocastor coypus	x	92	1	-	-
Reithrodon auritus	x	-	-	-	-
Calomys laucha	x	-	-	-	-
Mus musculus	-	-	1	-	-
Lagostomus maximus	x	-	1	-	-
Chinchillidae indet.	-	-	1	-	-
Zaedyus pichiy	-	-	-	7	1
Chaetophractus villosus	x	-	2	5	1
Dasypus hybridus	x	-	1	2	-
Tolypeutes matacus	x	-	-	-	-
+Scelidotheriinae	-	-	1	-	-
Lama guanicoe	x	1	9	4	5
Ozotoceros bezoarticus	x	5	2	4	2
Blastocerus dichotomus	-	9	-	-	-
Bos taurus	-	1	-	-	-
Ovis aries	-	1	-	-	-
Equus caballus	-	-	2	-	-
Felis concolor	x	-	-	1	2
Felis onca	x	-	-	-	-
Canis (P.) gymnocercus	x	-	-	-	-
Canis (P.) griseus	x	-	-	-	-
Chrysocyon brachyurus	x	-	-	-	-
Conepatus chinga	x	-	-	-	-
cf. Tursiops truncatus	-	1	-	-	-

"x": presence
"-": absence
Numbers: MNI (minimum number of individuals)

Table 2. Species recovered in sites 5, 6, 7, 8, 9, 10, 11, 12, 13, and 14.

TAXON	IL 5	CH 6	LM 7	FN 8	CT 9	LCh 10	L1 11	ZS 12	C 13	A52 14
Diplodon paralellopidedon	x									
Diplodon charruanus	x									
Ampullaria insularum	x									
Adelomedon brasiliensis										x
Amiantis sp.										x
Pseudoplatystoma	x									
Pterodoras		x								
Pimelodus sp.	x									
Rhea americana		x	x					x		
Rheidae indet.				x						
Rynchotus rufescens				x						
Plegadis sp.		x								
Eudromia elegans										x
Tinamidae indet.				x						
Nothura maculosa										x
Anatidae indet.	x		x							
Athene cunicularia										x
Lutreolina crassicaudata	x	x								
Ltra platensis	x	x								
Canis (P.) gymnocercus								x	x	
Canis (P.) culpaeus									x	
Canis (D.) avus?					x		x	x		
Lyncodon patagonicus										x
Canidae indet.				x						
Conepatus cf. castaneus					x					
Felis cf. colocolo					x					
Felidae indet.				x						
Cavia aperea	x	x			x		x			
Dolichotis patagonum							x	x		x
Lagostomus maximus	x			x	x		x	x		x
Hydrochoerus hydrochaeris	x	x								
Myocastor coypus	x	x	x	x						
Ctenomys sp.		x		x	x				x	x
Galea musteloides					x		x			

was under a moister and temperate cycle. It could be proved because of the faunal association (Table 1) and the palynological analysis (Salemme et al., i.p.; Salemme, 1987; Rabassa et al., i.p.).
5) Lechiguanas Island is an open-air site located in the Delta sector of the Paraná river, in the northeasternmost part of the Buenos Aires province (Caggiano, 1977). The older occupation is characterized by the absence of pottery and the

Table 2. Species recovered in sites 5, 6, 7, 8, 9, 10, 11, 12, 13, and 14.

TAXON	IL 5	CH 6	LM 7	FN 8	CT 9	LCh 10	L1 11	ZS 12	C 13	AS2 14
Cricetidae indet.				x						
Reithrodon auritus					x					
Akodon azarae					x					
Holochilus brasiliensis			x		x					
Cabreramys obscurus					x					
Calomys musculinus										
Oryzomys nigripes					x					
Oxymicterus rutilans					x					
Chaetophractus villosus			x		x		x	x	x	x
Dasypus septemcinctus			x							
Dasypus hybridus					x		x			
Zaedyus pichiy					x				x	x
Tolypeutes sp.									x	
Tayassu tajacu	x									
Lama guanicoe	x	x	x	x	x		x	x	x	x
Ozotoceros bezoarticus		x		x	x		x	x	x	x
Blastocerus dichotomus	x	x								
Ovis aries				x						
Bos taurus				x						
Equus caballus				x						
Monodelphis dimidiata					x					
Lestodelphis halli					x					
Didelphidae indet.			x							
Chiroptera indet.					x					
+Eutatus seguini					x	x				x
+Glyptodon sp.			x							x
+cf. Mylodon sp.			x							x
+Mylodontinae indet.			x							
+Glossotherium (G.) robustum										x
+cf. Hemiauchenia patachonica										x
+Hemiauchenia cf. wedelli										::
+Equus cf. E. (Amerhippus)										x
+Hippidion - Onohippidium										x
+Doedicurus clavicaudatus			x							
+Sclerocalyptus sp.			x							

presence of bone projectile points. The faunal remains correspond mainly to subtropical species but a few bones assigned to Lama sp. were found associated (Table 2). Radiocarbon dates place this episode in the beginnings of the Late Holocene (2,200 yr B.P., Table 3). Later occupations lacks of guanaco remains and pottery fragments are associated with subtropical species remains and mollusk shells.
6) Cañada Honda is a stratified site localized

along the homonymous creek, that was inhabited during the Late Holocene (Table 4); lithic implements, a great amount of pottery fragments and abundant faunal remains were associated. Also, a human skeleton with funerary goods accompanying the burial was found. The species represented in the site are, most of them, characteristic of the Subtropical Dominion, like Swamp deer (Blastocerus dichotomus), "capibara" (Hydrochoerus hydrochaeris), "coipo" (Myocastor coypus); but a few remains of guanaco (Lama guanicoe) were also found (Table 2), some of them decorated with geometrical drawings made by incision (Bonaparte, 1951).

7) La Moderna is an open-air site, in the margins of Azul creek, where two different human occupation succeeded. The former one could have been a glyptodon kill site because quartz instruments were associated with of Doedicurus clavicaudatus remains indicating that this big armadillo would have been swamped to be killed there (Politis, 1985). Besides this giant armadillo, the guanaco was present. A radiocarbon date for this Component (6,550 yr B.P., Table 3) has been considered as a minimum age. The Upper Component is characterized by quartz and quartzite instruments associated with guanaco and deer bones. That means that the earliest inhabitants used, at least occasionally, some of the pleistocenic megamammals still living at the beginning of the Holocene and then, during the second settlement -a campsite-, guanaco and deer were the basis of the diet (see list of species in Table 2).

8) Fortin Necochea, situated near a pampean pond in the Interserrana area, is an open air campsite, where several occupational episodes were detected. The latest one has been proposed as a posthispanic occupation in the XIXth century; thus, cultural materials and faunal remains reveal the hispanic influence . Cow, horse and sheep bones associated with guanaco and fossorial mammal remains were recorded (Crivelli et al., 1987). Before the Spaniards´ arrival, the site was inhabited by a hunter-gatherer group using quartzite instruments; they lived on the guanaco and Pampean deer hunting; although this occupation took place during the Holocene, this level was not radiocarbon dated; however, two radiocarbon dates performed on guanaco bones, coming from cultural lower levels still incompletely described, place earlier occupations

between 6,300 and 3,060 yr B.P.(Table 3). (Crivelli et al., 1985).

9) Tixi Cave in the hilly area of Tandilia, is a small rockshelter which was occupied temporarily by hunters. An assemblage of quartzite instruments and abundant mammal bones were recorded by Arana and Mazanti (pers.comm.). The faunal analysis revealed good climatic indicators, such as several species of rodents (see Table 2) and only a few bones of megamammals like guanaco and deer. Probably these last ones were used by the aborigines. The stratigraphic distribution of the taxa in the archaeological levels led to separate two faunal groups that show changes in the faunal composition from the lower up to the upper levels, that is fauna related with arid environmental conditions was replaced slowly by fauna that indicates wetter conditions. Two experimental radiocarbon dates performed on carbonates and coming from the upper levels (1,730; 1,220 yr B.P., Table 3) would indicate that the shelter was inhabited, at least, during the Late Holocene, and the fauna registered shows a period of transition from the cold, arid-semiarid conditions of the Holocene to a cycle more temperate and wetter. In fact, species characteristic of the Central Dominion were replaced by subtropical species (Table 2), which are moving from the north of Buenos Aires province (Tonni et al, 1988; Salemme, 1987). However, remains of an extinct armadillo were recorded in the lower levels in sediments assigned probably to the Late Pleistocene or Early Holocene, though the human occupation was not proved.

10) Cerro La China Locality, where three archaeological sites were found in small rockshelters, is located approximately 60 km farther northwest from Tixi Cave, in the Tandilia hilly area. In site 1 two archaeological levels were identified; the lower was dated at 10,730 +/- 150 and 10,790 +/- 120 yrs B.P. (Table 3) and included a preform of Fell's Cave Stemmed point and a fragment of a stem (Flegenheimer, 1980, 1987); no faunal remains were found there, except a scute of Eutatus sp., an extant armadillo which was deposited in related levels with those containing the projectile points. The upper level yielded triangular stemless points, pottery fragments and glass beads; there is no radiocarbon dating for this level (Zárate and Flegenheimer, i.p.).

Table 3. Chronological situation of the sites dated by C14.

PERIOD	C14 DATES	NORTH AREA	TANDILIA HILLY AREA	INTERSERRANA AREA	VENTANIA HILLY AREA
Historical times	440 ± 60		Loberia 1		
Late Holocene	.995 ± 65				La Toma CS
	1150 ± 70				La Toma 1
	1185 ± 35				La Toma 1
	1220 ± 70		Cueva Tixi		
	1560 ± 70				La Toma 2
	1730 ± 80		Cueva Tixi		
	2075 ± 70				La Toma 1
	2240 ± 55				La Toma 1
	2550 ± 90	Lechiguanas Island			
	2740 ± 80	Lechiguanas Island			
Middle Holocene	3630 ± 60			Fortin Necochea 2	
	3895 ± 100				La Toma 1
	5505 ± 200				La Toma 1
	5700 ± 120			Arroyo Seco	
	5740 ± 120			Arroyo Seco	
	6010 ± 400			Fortin Necochea 2	
	6550 ± 160			La Moderna	
Early Holocene	8390 ± 140			Arroyo Seco	
	8558 ± 316			Arroyo Seco	
	9780 ± 140				La Toma 2
Late Pleistocene	10720 ± 150	Cerro La China			
	10790 ± 120	Cerro La China			

11) Another rockshelter in the same area, 50 km to the southeast of the previous one, named Loberia 1 presents an abundant faunal assemblage (see Table 2), showing different environmental characteristics through the holocenic sequence; in fact, the lower levels include quartzite flakes associated with species from the Central and Patagonian Dominion and the upper levels yield fauna similar to the present one in the area, with a clear subtropical influence, although the guanaco would have been still present, probably. Numerous small triangular projectile points were recovered associated with the bones (Ceresole y Slavsky, 1985). A radiocarbon date place this latter human episode close to the Spanish Conquest (440 +/- 60

yr B.P., Table 3) and it is supposed that the
guanaco would have been refuged in the "sierras"
and would have constituted the main aboriginal
food. The other faunal remains found correspond to
rodents and dasypodids, that could be used as
complementary resources, but probably they were
deposited by other factors. Both micromammals
indicate, as in Tixi Cave, a transitional moment in
the environmental conditions.

12) Five sites in Zanjón Seco Locality yielded
abundant information about the human occupations
during the Late Holocene in the Interserrana area
(Politis y Tonni, 1982, 1985; Politis, 1984). From
all of them, lithic instruments, pottery fragments,
grinding stones and faunal remains were recovered,
both surficially and from stratigraphic positions.
Site 3 represents a place where the feeding wastes
were thrown and then approximately 3.000 bone
remains were recovered, including mainly guanaco
bones and secondarily Pampean deer bones; only a
flake was deposited there. Unfortunately, there are
no radiocarbon dates for any of these occupations
(Table 4).

13) Cortaderas is an unicomponent open-air site,
situated close to a creek. Faunal remains, lithic
artifacts and a few fragments of pottery were
associated. The bones corresponded to species
coming from the Central and Patagonian Dominion and
indicate clearly an open arid-semiarid environment
(Politis et al, 1983; see also Table 2).
Unfortunately, there are no published data about
cultural materials and none radiocarbon dates are
available for this site. Anyway, it could be
assigned to the Late Holocene, according to its
stratigraphic position and the presence of the
ceramic (Politis, 1984; see Table 4).

14) Arroyo Seco S2 is a campsite between a creek
and a presently dry lake, which was occupied
three times during the Holocene. The lower
Component is an assemblage of lithic artifacts,
like scrapers, flakes with retouched edges, a half
bola stone, knives, and bone remains of
extinguished species as well as extant species (see
Table 2). Likewise, seventeen human skeletons were
recovered below this assemblage; there were single
burials and communal burials, some of them with
funerary goods associated, such as marine shells,
glyptodon scute, ochre powder over the cranium.
This component was radiocarbon dated at 8,550 and
8,390 yr B.P. (Politis et al, 1987, see Table 3).

The Middle Component contained medium triangular stemmless projectile points, a portion of a grinding stone and quartzite flakes associated with bones remains of living species, like guanaco, Pampean deer, the autochtonous hare and armadillos. No radiocarbon dates exists for this occupation but it was assigned to the Middle Holocene (Table 4). Finally, the Upper Component is located in a ploughzone and some aboriginal materials (flakes, pottery, bone remains) are mixed with European material (glass, pieces of iron, but no bone remains corresponding to exotic species were found). Probably this episode took place just before the Spaniard´s arrival in the area, and the mixture would be due to the present anthropic activity (see Fidalgo et al, 1986; Salemme, 1987; Table 4).
After this brief summary, a series of hypotheses can be formulated, which are developed below.

DISCUSSION

According to the evidence, two groups of hypotheses have been formulated. First, the hypotheses related to the palaeoenvironments in Buenos Aires province (Pampean Region) during the Holocene; second, the adaptative strategies of hunter-gatherers in the area. In fact, the interpretation of geological aspects and the presence of faunal species led us to infer the environmental conditions to which the aborigines were adapted in that period. Thus, the following hypotheses have been proposed:
H 1: During the Early Holocene (10,000 to 6,000 yr B.P.) and Middle Holocene (6,000 to 2,500 yr B.P.), a big portion of the Buenos Aires territory was under arid to semiarid environmental conditions; the faunal species were characteristic from open areas and the more frequent correspond to those living at present in the Central and Patagonian Dominions. They were Dolichotis patagonum (the autochtonous hare), Lyncodon patagonicus (ferret), Lama guanicoe (guanaco), Zaedyus pichiy (piche, an armadillo) and Canis (P.) griseus (grey fox). Remains of these species were found within eolian and fluvial sediments which indicate colder and drier climatic conditions than the present ones in the area.
Remains of pleistocenic megamammals such as Equus cf. E. (Amerhippus) (an equid), Megatherium

Table 4. Chronological situation of the sites without absolute dates

PERIOD	NORTH AREA	TANDILIA HILLY AREA	INTERSERRANA AREA	VENTANIA HILLY AREA
Historical times			Fortin Necochea	
Late Holocene (2500 B.P.- 1500 A.D.)	Sitio Rio Lujan Canada Honda Canada de Rocha		Zanjon Seco 2 y 3 Cortaderas Tres Reyes 1	
Middle Holocene (6000 B.P.- 2500 B.P.)			La Moderna Arroyo Seco Tres Reyes	La Toma CI
Early Holocene (10000 B.P.- 6000 B.P.)				
Late Pleistocene				

americanum and Mylodon sp (giant sloths), Glyptodon sp., Eutatus sp. and Doedicurus clavicaudatus (extinct armadillos), Macrauchenia patachonica (a notoungulate) and Canis (Dusicyon) avus were found associated with cultural elements in several sites from the Interserrana area, like Arroyo Seco S2 (Fidalgo et al., 1986; Politis et al., 1987), and La Moderna (Politis, 1985) and from the Tandilia hilly area, like Cerro La China (Flegenheimer, 1980, 1987). Radiocarbon dates obtained from these sites roughly go from the Pleistocene-Holocene boundary (around 10,000 yr B.P.) into the Early Holocene.

Certainly, the "guanaco" is the species always present in the sites assigned to the Early and Middle Holocene. That is not only because of the environmental conditions but also, and mainly because of the anthropic activity (selectivity) and adjustment between the hunters and this prey .

H 2: Towards the Late Holocene, approximately 2,500 years B.P., the climatic and environmental conditions would have started to change from cold, arid-semiarid cycles to temperate, wet cycles similar to the present one. So the record in the palaeontological and archaeological sites shows some differences in the presence and frequency of the species. This is easily proved, specially in sites from the northeastern of Buenos Aires province, where the proportion of guanaco remains decreased and other elements from the Subtropical Dominion are registered, such as **Blastocerus dichotomus** (the swamp deer), **Hydrochoerus hydrochaeris** (capibara, a giant rodent), **Cavia aperea** (cuis), **Dasypus hybridus** (mulita, an armadillo) and **Myocastor coypus** (coipo). In the case of the coipo, it is necessary to point out that the essential requirement is a close water body because it is an ubiquitous species. However, important differences respect to its frequency in the record were detected in two sites very close each other, located in northeastern Buenos Aires province. In that sense, the remains of **Myocastor coypus** are very scarce in Cañada de Rocha meanwhile they are abundant in Río Luján Site. This fact could be an indicator of different conditions of humidity in an environment that today is very similar in both sites; moreover, it could be pointing out occupational episodes during different times.

The great abundance of fish remains in the records from the northernmost sites, such as Río Luján Site, Lechiguanas Island and Cañada Honda, is remarkable. Most of them are burned. However, scarce fish remains were present in Cañada de Rocha. So, alternative hypotheses about the high or low frequency and the use of fish species in those sites could be formulated: a) the fish richness was changeable in different epochs, although this is impossible to prove; b) a low frequency of megamammals (guanaco, swamp deer, pampean deer) during certain times, so the exploitation of fishes increased; c) distinct cultural customs in the feeding habits, which implied the use of other faunal resources.

In the sites of the hilly areas of Ventania and Tandilia a transitional fauna is observed, in which species from the subtropical and Patagonian environments are associated. That is the case of Tixi Cave (upper levels 1730 ; 1220 yrs. B.P.),

Lobería 1 cave (440 yrs B.P.) and La Toma, Upper Component (995 yrs B.P.). Both, the first and the second sites showed a great faunal diversity, mainly of rodents deposited by natural agents, which are very good climatic indicators. On the other hand, La Toma is an open-air site and so the megamammals are more frequent. Anyway, the association of "mulita" and "cuis" in La Toma and the rodents contexts in the caves, indicate a transitional environment.

In terms of chronology, some sites of the northern areas as Arroyo Frías and Cañada de Rocha could be assigned to the beginnings of the Late Holocene, because of the presence of species from the central environments, like "guanaco", Patagonian fox and the autochtonous hare. But other sites as Cañada Honda, Río Luján Site and Lechiguanas Island presented subtropical species, though there was scarce records of guanaco in their sequencies. If the radiocarbon date of Lechiguanas Island first occupation (2.220 yrs B.P.) where guanaco remains were found is considered, it seems that the environmental conditions were in transition to wetter cycles and some species were going towards the south and the west and some others were moving south of the Subtropical Dominion (see Table 5). Thus, all of these sites were inhabited during the Late Holocene, though under different environmental conditions, that is in two different moments, at least (see H3 below and Table 4).

According to this environmental scheme, several adaptive strategies are proposed; they are referred to the most useful resources and to the way they were employed.

H 3: The "guanaco" was the main resource during the Holocene. In Arroyo Seco S2 Lower Component, this camelid represents 50% of the record; in Zanjón Seco S3 comprises 99%; in Zanjón Seco S2, 80%; in La Toma, 60% and in Cortaderas, 80%. In that sense, Politis (1984) has proposed that the use of the "guanaco" can be considered as the basic annual resource for hunters living in the Interserrana area, at least for the occupational episodes registered. This affirmation can be extended to other areas of the Pampean Region, according to the data from the sites assigned to the beginnings of the Late Holocene in the Northern area and the corresponding ones to the hilly areas of Ventania and Tandilia.

The Pampean deer was a complementary resource in the aboriginal diet and in a similar way the armadillos (piche and peludo) and the flightless bird "ñandú" were used. These resources would have been available the entire year around, but the guanaco was, obviously, the prey of best yield. Although the meat of the ñandú would have been incorporated, the eggs were the most appreciable part of it; they were available from the end of winter and during spring and summer. Actually, abundant fragments of the shell eggs but only a few bone fragments of this species were found at the sites.

Extinct megamammals were an useful resource during the Late Pleistocene and Early Holocene, but only as occasional and complementary resources. Remains of several extinct species were found in Arroyo Seco S2, associated as feeding wastes and as funerary goods in human burials. Some other remains were recorded in Cerro La China Site and La Moderna Lower Component. These three sites have been radiocarbon dated (Table 3).

The records indicate at least two different episodes of occupation for the Northern area during the Holocene: 1) Arroyo Frías and Cañada de Rocha sites, which present the same characteristics of the Interserrana and hilly areas sites respect to the frequency of guanaco, Pampean deer and ñandú as the main resource the first one, and as complementary resource the others. They would have been inhabited during the earliest Late Holocene.
2) In other sites from the northeastern part of the territory (Cañada Honda, Lechiguanas Island and Río Luján Site), the Swamp deer and the coipo have been the main resource, whereas Pampean deer and fishes were complementary resources. Because of their stratigraphic position, the presence of pottery and an elaborated bone industry, these sites could be assigned to the rest of the Late Holocene (Table 6).

Meanwhile, in the hilly areas of Ventania and Tandilia and in the Interserrana area, the environmental conditions during the Late Holocene were changing, but the guanaco was present and it was abundant yet. This affirmation may also be verified with the late records of the sites Tres Reyes 1, Fortín Necochea, La Toma, Cueva Tixi, Lobería 1, Zanjón Seco 2 and 3 . In all of them, the camelid has been the main resource for the aborigines during the Late Holocene.

Table 5. Corology of the species registered at the
Holocenic sites.

PERIOD	NORTH AREA	INTERSERRANA AREA	HILLY AREAS
LATE HOLOCENE (2500 B.P. 1500 A.D:)	Swamp deer Pampean deer Otter Mulita Rea	Guanaco Pampean deer Peludo (armadillo) Mulita (armadillo) Rea Tuco-tuco (rodent)	Guanaco Pampean deer Peludo Mulita Rea
MIDDLE HOLOCENE (6000 B.P. 2500 B.P.)		Guanaco Pampean deer Peludo Piche (armadillo) Vizcacha (rodent) Grey fox Tuco-tuco Rea	Guanaco Pampean deer Peludo Piche Small rodent Rea Tuco-tuco
EARLY HOLOCENE (10000 B.P. 6000 B.P.)		Guanaco Pampean deer Piche Peludo Grey fox Pleistocenic megamammals Rea	Guanaco Pampean deer Piche Peludo Small rodent Pleistocenic megamammals

As proposed by Tonni and Politis (1980), the
guanaco disappeared completely from the Buenos
Aires province by the time of the Spanish Conquest;
none of the chronicles of the XVI and XVII
centuries mentions the presence of guanacos in the
Northern and Interserrana areas. Thus, they tell
frequently the hunt of Pampean deers and ñandúes in
both areas, where the first of the species would
have been the main resource during the historical
times, probably due to the easy catch and its major
availability.
However, several populations of guanaco could have
been remained isolated in the Tandilia hilly area,
which had been acted as a refuge until historical
times, as it was suggested in the records of
Lobería 1 and Tixi Cave sites. In the Ventania
hilly area, troops og guanaco can be still found
today, but they are composed by a few individuals.
Anyway, it is remarkable that, from a
zoogeographical point of view, the area is an
"enclave" from the Central Dominion in the Pampean
Dominion (Ringuelet, 1955; 1961).

Likewise, according to the cites of the chroniclers, the introduction of the European horse in the culture would have modified the food preferences. In fact, this is evidenced through the comments of the voyagers of the XVII and XVIII centuries.

Table 6 summarizes the evidence proposed for H3 and then a general hypothesis can be formulated, based in the previous H1, H2 and H3:

H : The boundaries of the Pampean Dominion have changed back and forth in several occasions during the Holocene. In fact, during the Early and Middle Holocene, some species from the Patagonian and Central Dominion migrated northwards and eastwards of their present habitats. On the other hand, since 2,500 yrs. B.P. and during the Late Holocene, some species from the Subtropical Dominion began to move southwards, yielding the retraction of the Patagonian and Central species to their own present environments. Finally, at present, the Pampean Dominion is clearly influenced by Brazilian species of the Subtropical Dominion.

In the second group of hypothesis about the faunal exploitation and the aboriginal adaptation to the environment, the following ideas were formulated:

H 4: The aborigines of the Buenos Aires sector of the Pampean Region were hunter-gatherers who selected their preys from the species available during the entire year. The hunting techniques employed could have been similar to those suggested from the ethnohistorical information. Communal hunting could have been employed for the guanaco, ñandú and probably for the swamp deer, whereas the Pampean deer would have been hunted by men alone. Probably, both techniques were used frequently by the men of those bands; however, some other species like rodents (coipo, vizcacha) and armadillos could have been hunted by anybody within the band, even the children (Politis and Salemme, i.p.).

The cooperative hunting technique could have been applied also to the Pleistocenic megamammals, in the same way for those of lonely habits (D. clavicaudatus, Glyptodon sp., Mylodon sp) as those that used to live in herds (Equus [Amerhippus]). In the case of the sloths and giant armadillos, they would have been swamped to kill them.

Table 6. Available resources during the Holocene

AVAILABLE RESOURCE	NORTH AREA				INTERSERRANA AREA				HILLY AREAS			
	H1	H2	H3	Th	H1	H2	H3	Th	H1	H2	H3	Th
Guanaco	-	-	C	-	M	M	M	C	-	M	M	C
Pampean deer	-	-	M	-	C	C	C	M	-	C	C	P
Swamp deer	-	-	M	M	-	-	-	-	-	-	-	-
Piche-Peludo	-	-	-	-	C	C	C	C	-	C	C	C
Mulita	-	-	C	-	-	-	-	-	-	-	C	-
Otter	-	-	M	C	-	-	-	-	-	-	-	-
Rea	-	-	C	C	C	C	C	C	-	C	C	C
European horse	-	-	-	-	-	-	-	C	-	-	-	C
Pleistocenic megamammals	-	-	-	-	C	-	-	-	C	-	-	-

References
M: main resource
C: complementary/occasional resource
-: without data
H1: Early Holocene
H2: Middle Holocene
H3: Late Holocene
Th: Historical times

        A new technique for hunting the guanacos was
employed after the introduction of the European
horse; it was hunted by men alone, riding horses.
This assertion is valid for the Historic Tehuelches
in northern Patagonia, but probably the same
technique had been used in the south of Buenos
Aires province at that time, where some guanacos
could have remained there, as they do that today.
        The fishing techniques used by the aborigines
are still unknown, but probably the utilization of
harpoons from the river banks would have been the
more efficient method to catch any of the species
of Doradidae. Several species of this family lived,
and are living today, in the northeastern streams
of Buenos Aires province; commonly, they are
swimming on the surface, so that the fishing with
harpoons probably was the more frequent technique
employed. The other possibility could be the
fishing from canoes, but there is neither
archaeological nor ethnohistorical evidence
available.
        It is interesting to point that the only sites
where fish remains have been recovered were those
localized in the northeastern sector of the studied

region, even though the other sites were located close to body waters as well. The aforementioned sites are Cañada de Rocha (although fish remains are not abundant), Sitio Río Luján, Cañada Honda and Lechiguanas Island.

H 5: The Pampean aboriginal diet based on animal proteins would have been complemented not only with the gathering of ñandú eggs, available during spring/summer and easy to get, but also with products of some vegetable species. These could be the fruits of "algarrobo" and "chañar", both species living today in the phytogeographical Provincia del Espinal, but they could have invaded the Pampean Dominion during the arid or semiarid Holocene phases (Frenguelli, 1941) and possibly during the Late Pleistocene. Some chroniclers mention in their comments the presence of several xerophitic and thorny species surveyed during their travels in the XVII and XVIII centuries; other evidences were recognized in the Sierra de la Ventana area (southwestern of Buenos Aires province) (J.Frangi, pers. comm.).
    The gathering had to be performed mainly during the spring and summer, because the fruits of these species are available in these seasons. There is no direct evidence of gathering, but some implements like mortars, pestles and "manos" suggest that their main function was the grinding of grains and fruits.

H 6: The megamammals were used wholly, i.e. meat and fat for feeding, leather and skin for clothes and tents, bones as raw material to make instruments and ornaments (Salemme and Miotti, 1987), and occasionally, the blood was employed for drinking (Schmidel, 1969).
    As it was previously mentioned, the guanaco has been the main resource and the Pampean deer was a complementary resource. Although both of them were available all year around, the bigger size, the social behaviour and the well-known territoriality of guanaco would have been the principal fact of adaptation between the prey and its predator. In fact, the MNI tables of the studied sites (Salemme, 1987) show that Lama guanicoe is predominant compared with Ozotoceros bezoarticus, except in those sites of the northeastern area where the guanaco was scarce during the Late Holocene, thus been later replaced

by Blastocerus dichotomus. Anyway, the recorded frequencies of both cervids are similar.

In summary, the Buenos Aires sector of the Pampean Region had been inhabited by hunters, probably organized in bands, at least since the Early Holocene. They were efficiently adapted to the changing environmental conditions that took place during the last 10.000 years. The climatic oscillations occurring since the Early Holocene until historical times have influenced the corology of several mammal species and, likewise, the human group adaptation to those changing conditions.

In a smaller scale, it is possible to detect modifications in the distribution of some species in the Pampean Dominion from the Northeast to the Southwest of it and since the Early Holocene to the historical times.

It must be emphazised that since 1.500 A.D., exotic species brought by the Spaniards, like horse, cow and sheep were added to the autochtonous fauna. Those introduced species modified the aboriginal adaptive systems. Remarkable changes in the economic structure of the hunters produced a sudden transculturation, which ended shortly with the absortion of the aborigines and their extermination.

REFERENCES

Ameghino,F. 1880.. La antiguedad del Hombre en el Plata. Impr. Coni 2 Vol., 1000 pags. Bs.As.

Bonaparte,J. 1951. Nota preliminar de un paradero aborigen en la Cañada Honda (partido de Baradero). Arqueología 2. Mus. Cs. Nat. Carlos Ameghino. Mercedes (B).

Cabrera, A.L. 1971. Fitogeografía de la República Argentina. Bol.Soc.Arg. Botanica, XIV (1-2): 1-42

Caggiano, M.A. 1977. Contribución a la arqueología del Delta del Paraná. Obra Centenario Mus.La Plata, Antrop., II: 301-324.

Ceresole,G. & L. Slavsky 1985. Informe preliminar sobre la Localidad Lobería 1 (provincia de Buenos Aires). VIII Congr.Nac.Arqueol.Arg., resumen: 4. Concordia.

Crivelli,E.; M. Silveira; E. Eugenio; P. Escola; M. Fernández & N. Franco. 1987/1988. El sitio Fortín Necochea (Pdo. de Gral. Lamadrid, Pcia. de Buenos Aires). Estado actual de los trabajos. Paletnologica,1: 7-37. Buenos Aires.

Fidalgo,F.; L.Meo Guzmán; G. Politis, M. Salemme &
E. Tonni. 1986. Investigaciones arqueológicas en
el sitio 2 de Arroyo Seco (Pdo. de Tres Arroyos,
Pcia. de Buenos Aires, Rep.Argentina). **New
Evidence for the Peopling of the Americas,**
A.Bryan (Ed.), pag.: 221-269. Univ. of Alberta,
Canada.

Flegenheimer,N. 1980. Hallazgos de puntas "cola de
pescado" en la provincia de Buenos Aires.
Rev.Relaciones, Soc.Arg.Antrop., XIV (1): 169-176.

Flegenheimer,N. 1987. Recent research at Localities
Cerro La China and Cerro El Sombrero. **Current
Research in the Pleistocene,** Vol.4. Maine.

Frenguelli,J. 1941 Rasgos principales de la
fitogeografía argentina. **Rev.Mus.La Plata,** III,
Botánica: 65-181.

Madrid,P. & Salemme,M.1988. La ocupación tardía del
Sitio 1 de la Laguna Tres Reyes (Pdo. de González
Chaves, Pcia. de Buenos Aires). **IX
Congr.Arqueol.Arg.,** abstract: 57. Buenos Aires,
november 1988.

Petrocelli, J.1975. Nota preliminar sobre hallazgos
arqueológicos en el valle del río Luján
(Población Río Luján, Campana, Pcia. de Buenos
Aires). **Actas I Cong.Arq.Arg.:** 251-2700.
Rosario.(1970).

Politis,G. 1984. Arqueología del área Interserrana
bonaerense. **Tesis doctoral,** Fac.Cs.Nat. y Museo,
397 pags.

Politis,G. 1985. Hombre temprano y gliptodóntidos
en la Región Pampeana (Argentina): el sitio de
caza "La Moderna". **Simp.Cazad.-Recolect. de
America,** 45 Cong.Intern.Americ., Bogotá,
Colombia.

Politis,G. & Madrid,M. 1988. Un hueso duro de roer:
análisis preliminar de la tafonomía del sitio
Laguna Tres Reyes 1 (partido de Gonzales Chaves,
Pcia. de Buenos Aires). **Semin.Arqueofauna de
Vertebrados e Invertebrados de sitios
arqueológicos,** Fac. Fil. y Letras, Buenos Aires,
12 pags.

Politis,G. & E. Tonni. 1982. Arqueología de la
Región Pampeana: el sitio 2 de Zanjón Seco
(Distrito de Necochea, Pcia. de Buenos Aires,
Rep.Argentina). **Rev.Pre-Historia,** Vol. III (4):
109-142, Univ. de São Paulo, Brasil.

Politis,G. & E. Tonni. 1985. Investigaciones
arqueológicas en el sitio 3 de Zanjón Seco,
partido de Necochea. **1ras.
Jorn.Antrop.Pcia.Bs.As.,** Chivilcoy.

Politis,G. & M. Salemme. in press. Prehispanic mammal exploitation and hunting strategies in the Eastern Pampa Subregion of Argentina. **Symp.on Communal Land Mammal Hunting and Butchering.** The World Archaeol. Congr., Oxbow Books, London.

Politis,G.; E. Tonni & F. Fidalgo. 1983. Cambios corológicos de algunos mamíferos en el área Interserrana de la provincia de Buenos Aires durante el Holoceno. **Ameghiniana,** 20 (1-2): 72-80.

Politis,G.; E. Tonni; F.Fidalgo; M. Salemme & L. Meo Guzmán.1987. Man and Pleistocene megamammals in the Argentine Pampa: Site 2 at Arroyo Seco. **Current Research in the Pleistocene,** Vol.4:159-161. Center for the study of Early Man, Univ. of Orono

Rabassa,J.; A.Brandani; M. Salemme & G.Politis.in press. La Pequeña Edad del Hielo" (siglos XVII a XIX) y su posible influencia en la aridización de áreas marginales de la Pampa Húmeda (Pcia. de Buenos Aires). **Actas I Jorn.Geol.Bonaer.** Tandil. 1985.

Ringuelet,R. 1955. Panorama zoogeográfico de la provincia de Buenos Aires. **Notas Pelim.Mus.La Plata,** Zool. 18 (156): 1-45.

Ringuelet,R. 1961. Rasgos fundamentales de la zoogeografía de la Argentina. **Physis** 22 (63): 151-170.

Salemme,M. 1983. Distribución de algunas especies de mamferos en el Noreste de la Provincia de Buenos Aires durante el Holoceno. **Ameghiniana,** 20 (1-2): 81-94.

Salemme,M. 1987. Paleoetnozoología del sector bonaerense de la Región Pampeana, con especial atención a los mamíferos. **Tesis Doctoral,** Fac.Cs.Nat., Univ.Nac.La Plata.267 pags. La Plata

Salemme,M. & E.Tonni. 1983.Paleoetnozoología de un sitio arqueológico en la Pampa Ondulada: Sitio Río Luján (Pdo. de Campana, Pcia. de Buenos Aires). **Rev. Relaciones,** Soc.Arg.Antrop., XV: 77-90.

Salemme,M. & L.Miotti. 1987. Zooarchaeology and palaeoenvironments:some examples from the Pampean and Patagonian Regions (Argentina). **Quat. of South Amer. & Antarct.Penin.,** Vol.5 : 33-57. Balkema Publ., Holland.

Salemme,M.; G.Politis; P. Madrid; F. Oliva & L. Guerci. in press. Informe preliminar sobre las investigaciones arqueológicas en el sitio La Toma, Pdo de Cnel. Pringles, Prov. de Buenos Aires. **VIII Congr.** Nac.Arqueol.Arg., Actas, Concordia, 1985.

Schmidel,U. 1969. Viaje al Río de La Plata y
   Paraguay. Col. Obras y Docum., P.de Angelis.
   T.VI: 245-346. Edit.Plus Ultra.
Tonni,E.; M.S. Bargo & J.L.Prado. 1988.Los cambios
   ambientales en el Pleistoceno tardío y Holoceno
   del sudeste de la provincia de Buenos Aires a
   través de una secuencia de mamíferos.
   Ameghiniana, 25 (3-4).